Bus-based
Industrial Process Control

Bus-based Industrial Process Control

Introducing the international STEbus standard

MICHAEL TOOLEY

Heinemann Newnes

Heinemann Newnes
An imprint of Heinemann Professional Publishing Ltd
Halley Court, Jordan Hill, Oxford OX2 8EJ

OXFORD LONDON MELBOURNE AUCKLAND SINGAPORE
IBADAN NAIROBI GABORONE KINGSTON

First published 1988
Reprinted 1988

© Michael Tooley 1988

British Library Cataloguing in Publication Data

Tooley, Michael
  Bus-based industrial process control:
  introducing the international STEbus
  standard.
  1. Industries. Process control. Applications
  of microcomputer systems
  I. Title
  670.42'7

ISBN 0 434 92009 6

Photoset by Deltatype Ltd, Ellesmere Port, Cheshire
Printed in Great Britain by Redwood Burn Ltd, Trowbridge, Wiltshire

Contents

Introduction		1
1	Introduction to bus-based systems	4
2	Introduction to microcomputers	22
3	STEbus hardware	48
4	Operating systems	79
5	Assembly language programming	107
6	BASIC programming	143
7	C programming	159
8	Interfacing	172
9	STEbus applications	213
10	Reliability and fault-finding	243
Appendices		
1	STE backplane signals	256
2	Decimal, binary, hexadecimal and ASCII conversion table	257
3	The STE Bus Product Guide	263
4	Commonly used STE digital I/O bus pin connections	264
5	Commonly used STE analogue I/O bus pin connections	265
6	IEEE-488 pin connections	266
7	RS-232C pin connections	268
8	The STE Manufacturers and Users Group	270
9	BASIC commands, compiler directives, functions and statements	271
10	BASIC-Z80 procedure library	294
11	An overview of STEbus from the STE Manufacturers and Users Group	296
Glossary of bus terms		307
Bibliography		311
Index		315

Introduction

The advent of powerful yet easy to use microcomputer systems has led to something of a revolution in the field of industrial process control. Such systems are invariably bus-based and are inherently flexible since functional modules can be assembled in a huge variety of different configurations.

Bus-based microcomputer systems are more powerful and more flexible than systems based on programmable logic controllers (PLC). They are also more flexible and significantly less expensive than systems based on personal computers.

The last decade has seen the emergence of a wide range of bus standards. Such standards relate not only to the physical characteristics of the backplane used to interconnect the various functional modules but also relates to the protocol used for exchanging data on the bus. The choice of bus standard is crucial.

The standard Eurocard bus is set to become the standard for small- and medium-scale industrial control, instrumentation and data logging microcomputer applications. The standard provides a simple yet highly flexible solution to a wide range of industrial process control, instrumentation and data logging problems.

The origins of the Standard Eurocard (STE) bus can be traced back to 1981 when GMT Electronics were seeking a Eurocard-based 8-bit backplane bus which would act as a vehicle from which to launch a new product range. The search for a suitable standard proved to be fruitless. However, while none of the existing bus systems possessed the desired attributes (many were severely restricted by virtue of their dependence on the control signals generated by a particular microprocessor) one of the systems considered by GMT stood out as being of particular interest. This standard, known as STD, was the subject of an IEEE working group. Accordingly, a proposal was made to extend the standard in order to encompass a Eurocard implementation. In retrospect, it is perhaps not surprising that this proposal was rejected. Despite this, the need for an 8-bit processor independent bus standard based on the Eurocard format was at last recognized and a further working group was convened to consider this new standard.

With the exception of GMT, none of the members of this new working group had any immediate commercial interest in the proposed standard. This fact was undoubtedly instrumental in the production of a specification which is truly processor independent and not over-complicated by the need to satisfy numerous criteria imposed by the need to retain compatability with existing bus implementations. The STEbus system emerged as a standard designed by engineers for engineers.

The first STE products became available in 1984. In the same year, the STE Manufacturers and Users Group (STEMUG) was formed. This body, more than any other, has been responsible for the promulgation of the STEbus and, furthermore, has been instrumental in making further minor technical improvements in the IEEE-1000 standard.

This book aims to provide readers with sufficient information to be able to select the necessary hardware and software to implement a wide range of industrial microcomputer systems based on the increasingly popular STEbus. The book contains numerous examples of bus hardware and software and two practical bus applications are discussed in some detail. Representative software has been included in assembly language, BASIC and C. Information has been included so that circuits and software routines can be readily modified and extended by readers to meet their own particular needs; the principal aim has been that of providing readers with sufficient information so that they can solve a wide variety of engineering problems in the shortest possible time and without recourse to any other texts. Furthermore, while the book has been aimed at the practising technician, engineer and process control specialist, it does not assume any previous knowledge of microprocessors or microcomputer bus systems. The book is therefore eminently suitable for those looking for new solutions to engineering problems.

Chapter 1 introduces the reader to bus-based systems and includes an overview of some of the most popular bus standards. The chapter also examines the desirable characteristics of a bus and shows how STE closely approaches the ideal as far as 8-bit operation is concerned.

For the benefit of those having no previous knowledge of microcomputers, Chapter 2 provides an introduction to microprocessors and microcomputer systems. The chapter also includes information on microprocessor support devices (including semiconductor memories and programmable I/O (input/output) devices and contains sections devoted to operating systems and programming languages.

Chapter 3 provides a thorough description of the STEbus and deals with functional elements, signals and arbitration. Some typical STE products in the shape of processor and I/O cards are also described.

Microcomputer operating systems are discussed in Chapter 4. The details and examples given relate to the ever-popular CP/M-Plus operating system. However, equivalent functions and commands are found in virtually every popular operating system. The aim of the chapter is that of outlining the facilities provided by an operating system which are of particular interest to the software developer.

Chapter 5 aims to provide readers with an overview of assembly language programming techniques and explores the architecture and instruction sets of two of the most popular microprocessors used in low-cost process control applications. The aim has been to provide readers with sufficient information to make an informed choice of processor type, to outline the advantages and disadvantages of assembly language programming and to introduce techniques used in developing assembly language programs.

Chapter 6 and 7 are devoted to programming in BASIC and C respectively. The essential features of each language are discussed and numerous examples are provided which may be readily incorporated in the reader's own programs.

Chapter 8 deals with the general principles of interfacing sensors and transducers to STEbus modules. A number of common sensors and transducers are described and, for those who do not wish to make use of 'off-the-shelf' signal conditioning modules, details the circuitry necessary to interface such devices to standard STE I/O cards.

Chapter 9 provides readers with an introduction to the procedure for selecting hardware and software for typical STEbus applications. In addition, two representative applications are discussed in depth and the hardware configuration and sample software is given for each.

Chapter 10 deals with quality and reliability in the context of bus-based systems. The chapter also examines some basic test equipment and fault-finding techniques applicable to bus-based systems.

The book concludes with an extensive reference section comprising eleven separate appendices, a glossary of bus terms and a bibliography.

Finally, I would like to extend my sincere thanks to members of the STE Manufacturers and Users Group and others who have helped in the production of this book. I would particularly like to thank David Page and Mohander Matu of RS Components Ltd for their invaluable help and for the kind loan of a multiplicity of STE bus hardware. Thanks also go to Alan Timmins of BICC-VERO Electronics, Steve Hinton of TRENAN, Bob Squirrell of STEMUG (and editor of the *STE Bus Product Guide*), Chris Jarvis of Farnell Electronic Components, and Tony Franklin of British Telecom, Barrie Nicholson of Wordsun, Anthony Winter of Arcom Control Systems, Charles Palmer of DSP Designs and Chris Roberts of Brooklands Technical College.

Mike Tooley

1
Introduction to bus-based systems

There is nothing particularly new about the concept of a bus-based control system. Indeed, some popular bus systems have been in regular use for over a decade. This chapter sets out to explain what a bus is and why it offers significant advantages over any other technique of assembling a microcomputer or microprocessor-based system. We shall also be examining some of the most popular and successful bus systems, ending with a bus system which promises to be *the* 'industry standard' bus for the vast majority of industrial process control applications.

Bus terminology

The word 'bus' is a contraction of 'omnibus', a Latin word which simply means 'to all'. This aptly and succinctly describes the function of a bus, i.e. the linking of components within a complex system by means of a common highway. Not only does such an arrangement minimize the total number of connecting paths but it also prevents duplication by allowing intelligent devices shared access to ancillary subsystems (such as memories and I/O cards).

In the context of microprocessors and microcomputers there are, of course, several types of bus. Very large scale integration (VLSI) devices (such as microprocessors and their associated support devices) invariably have an internal bus structure. This usually takes the form of a high speed internal data bus which is buffered from the outside world and links together all of the internal registers. To eliminate confusion we shall refer to this type of bus as an *internal bus*.

Another type of bus is that which links together various VLSI devices (such as microprocessor, memory and I/O devices) within a single printed circuit board (PCB) card. Here there will usually be three distinct buses present: a *data* bus to provide a path for data; an *address bus* to provide a path for addresses generated by the microprocessor; and a *control bus* to cope with the various control signals required. We shall refer to this bus structure as a *local bus*.

To minimize the number of connecting paths, the data and address buses within a local bus may make use of a common set of interconnections on a time shared basis. Such an arrangement is known as a *multiplexed address/data bus*.

Another form of bus is used to link together the subsystem components (each comprising a PCB card and a number of VLSI devices) within a more complex system. Such a system may incorporate several microprocessors (each with its own support devices and local bus system) or may just involve a single microprocessor (again with support devices and its own local bus) operating in conjunction with a number of less intelligent supporting cards.

This system generally comprises a number of cards mounted in a frame and linked together at the rear of the card frame by tracks on a printed wiring board mounted at right angles to the cards. Such an arrangement is more correctly known as a *backplane* bus and this book is devoted to such systems. Therefore, from this point on, we shall take the word bus to mean backplane bus unless otherwise qualified.

The origins of most modern microcomputer bus systems can be traced back to the early 1970s. The first users of bus systems were the major computer manufacturers who developed their own bus standards in order to allow a wide range of permutations within complex computer systems assembled using a number of subsystem components.

This concept was followed in the late 1970s by the major semiconductor manufacturers, each seeking a vehicle for promoting its own range of VLSI components. In this case the bus system was really nothing more than an elaborate extension of the basic microprocessor bus. Such systems were invariably synchronous (operating at speeds governed by the microprocessor clock) and, since only one processor was present within the system, there was no contention for the use of the bus.

Unfortunately, since each manufacturer was keen to promote the use of its own standard, a proliferation of bus standards appeared. This unhappy situation was further compounded as the move towards 16- and 32-bit processors together with the need for a bus to support several intelligent devices (rather than just one) meant that many of the original standards were developed and extended to the point of absurdity. Several systems 'grew like topsy' to become nightmares for the unfortunate end user beset with the ever pressing need to retain compatibility within standards which would just not remain static.

Advantages of bus systems

Bus systems are inherently flexible and present the designer with a huge range of options for optimizing system performance without affecting system compatibility. The modular nature of the bus allows a large number of permutations and systems can be readily expanded to cope with changing requirements and new demands.

The ability to keep pace with improvements in technology is becoming a vitally important consideration in the design of electronic systems generally. Indeed, with typical product development times of two to three years, we are rapidly approaching the point at which a product can be out of date before it goes into production.

The suspicion that a product may be out of date when it makes its first appearance in the marketplace can be a strong disincentive to those responsible for product development. The solution, of course, is simply that of designing with upgrading and enhancement in mind. The use of a bus system can be instrumental in ensuring that performance can keep pace with improvement in technology; the system can be progressively upgraded (with new cards added or old cards replaced) to maintain a standard of performance which is commensurate with 'state of the art' technology.

Relative immunity to inbuilt obsolesence is, however, not the only benefit associated with bus-based systems. Modularity allows exchange of modules for servicing; cards can be rapidly exchanged when faults develop or interchanged between racks for testing. The end-user is thus able to minimize system down-time and need not be concerned with board level servicing as cards can be returned to manufacturers or service agencies while the system remains fully operational.

Bus concepts

It is sometimes convenient to think of a bus as existing on a number of levels and, in particular, to make a distinction between the conceptual and logical aspects of a bus and its underlying physical structure.

A bus can thus be thought of as existing on various levels; the three most important of which may be described as conceptual, logical, and physical. At the conceptual level we are concerned with the basic philosophy of a bus and, in particular, the division of subsystems present into *masters* (intelligent controlling devices which can generate bus commands), *slaves* (devices which generally exhibit less intelligence and cannot themselves generate bus commands), and *intelligent slaves* (these are slaves which have their own intelligent controlling device but which do not themselves have the capacity to place commands on the bus).

The logical level is very much concerned with the passage of information in the form of signals on the bus. Here we are concerned with the protocol which must be adhered to for orderly exchange of data on the bus and the mechanisms for data transfer and the exchange of bus control.

Finally, the physical level is concerned with such mundane (but none the less important) items as the dimensions of the subrack, backplane, and cards, and the means of connection to the backplane.

Timing, sequence and flow diagrams

Several types of diagram are commonly used to explain the operation of a bus and the interchange of bus signals. Diagrams generally fall into one of three classes related to timing, sequence and flow control.

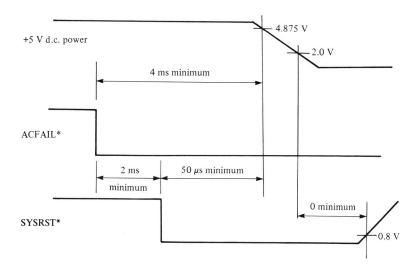

Figure 1 *Representative timing diagram which defines the response of a system to AC power failure*

A bus standard will usually specify maximum and/or minimum times for bus signal transitions. Such limits impose obvious constraints on the backplane interface logic and timing diagrams show these relationships in graphical form. Figure 1 shows a representative timing diagram in which the AC power failure response of a system is defined.

Sequence diagrams are concerned with the interlocked timing relationships of the functional modules present within a bus system. Here we are rather more concerned with a sequence of events than a set of timings. Figure 2 shows a typical sequence diagram in which module A cannot

8 Introduction to bus-based systems

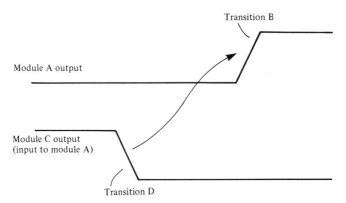

Figure 2 *Typical sequence diagram showing handshake protocol*

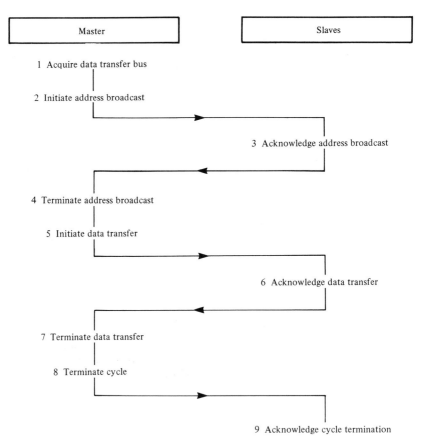

Figure 3 *Flow diagram illustrating a data transfer bus cycle*

generate transition B until it detects module C's generation of signal transition D. Timing and sequence diagrams may sometimes be usefully combined in which case time limits are simply shown superimposed on the sequence diagram.

Flow diagrams show a stream of events as they would occur during a bus operation. Events are stated in words and the diagram is presented in a sequential manner (invariably moving down the page). Such diagrams can be used to gain an impression of the interaction of various functional modules (e.g. a master and several slaves) within a bus. Figure 3 shows a representative flow diagram for a data transfer bus cycle.

Physical features of a bus

The essential features of a bus system are:

1. A data path (usually comprising 8, 16 or 32 lines).
2. A number of address lines for memory and I/O addresses.
3. Bus control lines (e.g. bus request and bus grant signals).
4. A bus clock will be required in the case of a synchronous bus.

Bus lines take the form of printed circuit tracks which link similarly numbered pins on a number of connectors soldered to the backplane printed circuit board. The subsystem cards are fitted into guide slots which align with the backplane connectors and are simply pushed home until a satisfactory mechanical and electrical connection is made.

To ensure high data rates the electrical length of the backplane should be kept as short as possible and the backplane should be terminated in order to minimize propagation delays and standardize the characteristics of the bus lines.

Asynchronous bus systems

Board-level systems use conventional synchronous data transfers and operation is governed by the locally generated microprocessor clock. This synchronous method of data transfer is, however, inappropriate in the case of a backplane bus where the system would be constrained to operate at a clock speed determined by the slowest microprocessor present.

A far better method involves asynchronous data transfers. This method relies on a handshaking protocol and data transfers take place at speeds which are determined by the participating devices.

Bus architecture

We have already mentioned that bus systems generally support three types of subsystem: masters, slaves, and intelligent slaves. A simple bus system may have only one master supported by a number of slaves, each performing some dedicated activity such as analogue I/O, digital I/O, or disk control. A simple single-processor bus system of this type is depicted in Figure 4. The master effectively has a monopoly of the slaves, accessing and controlling each as the need arises. The control of data transfers on the bus is thus vested in the processor. As an example, consider a simple data transfer operation using asynchronous master-slave handshaking protocol shown in the timing/sequence diagram of Figure 5.

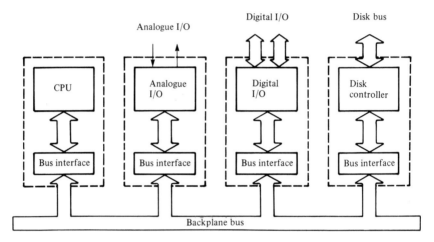

Figure 4 *Architecture of a simple single-processor bus system*

The processor (bus master) places the memory or I/O port address on the address bus. If the operation is a write, the data is also placed on the data lines. The bus master then generates a command (I/O read or write or memory read or write) which activates the appropriate bus slave. The slave accepts the data if the operation is a write, or places data on the bus if the operation is a read. A transfer acknowledge is then sent to the bus master by the slave. This allows the master to complete its cycle, removing the command from the bus and then removing the address and data from the bus interface.

The ability to support more than one master is clearly a useful facility and one which more fully exploits the potential of a bus system. Such an arrangement is known as a *multiprocessor system*.

While several bus masters may be connected in a bus configuration it should be clear that *only one* can command the bus at any time. The resources offered by the slaves must be shared between several bus masters. Each bus master must thus be considered as a *potential* master; it

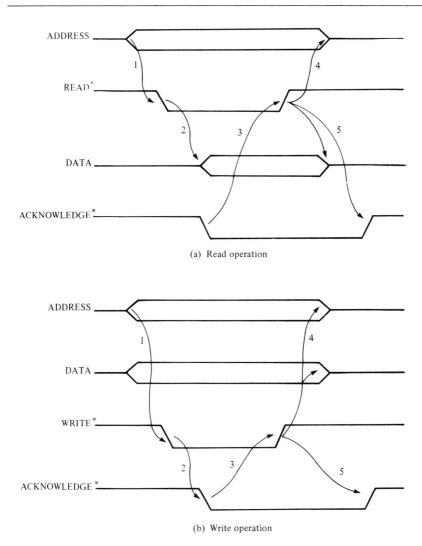

Figure 5 *Timing/sequence diagram showing asynchronous master-slave handshaking protocol*

cannot have exclusive use of the bus. Once access has been acquired by a master device, it has a period of *exclusive* control to carry out data transfers by means of a sequence of command signals, memory and I/O addresses. Once the process of transfer is complete the master *must* release the bus to allow other potential bus masters access or, depending upon protocol, may hold the bus until requested by another master.

All modern bus systems can support several bus masters. In order to eliminate possible contentions for the use of the bus a system of bus arbitration is required. Several techniques are used to establish priority and these generally fall into two classes: serial and parallel, as shown in Figures 6 and 7 respectively.

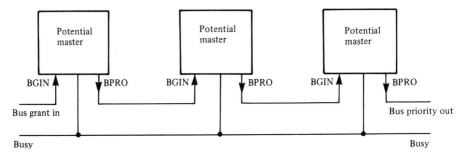

Figure 6 *Serial arbitrated system*

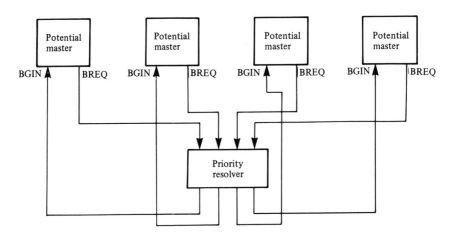

Figure 7 *Parallel arbitrated system*

In a serially arbitrated system, bus access is granted by a priority determined by the physical slot location. Each master present notifies the next lower priority master when it needs to gain access to the bus. It also monitors the bus request status of the next higher-priority master. The masters thus pass bus requests along from one to the next in a daisy chain fashion.

In a parallel arbitrated system, external hardware is used to establish the priority of each bus master. Both systems have their advantages and disadvantages and some bus standards permit the use of both techniques.

Interrupt handling

Any practical industrial process control system must have some provision for handling interrupts. These are signals generated by external devices which require the immediate attention of the control system. An obvious example of the use of interrupts is in the case of AC power failure. By taking immediate action when the AC mains drops out, it is possible for the system to take the necessary action in order to protect itself. Another example would be in the case of an emergency stop button; we would want the system to respond to this regardless of whatever else was happening at the time. Interrupt management is thus vitally important to the operation of a backplane bus system.

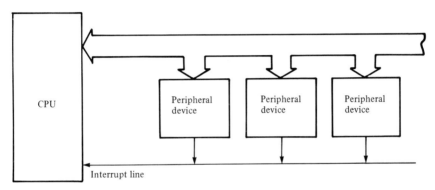

Figure 8 *Polled interrupt system*

Three distinct methods of handling interrupts are known as *polled*, *multiline* and *bus vectored*. Polled interrupt systems generally employ a single interrupt request line (as shown in Figure 8). An external subsystem signals its need for service by asserting the signal on this line. Provided that the processor is currently accepting interrupts (i.e. that its internal logic has not been placed in a state in which interrupts are 'masked'), the processor suspends its current operation (saving important data so that an orderly return can be made to the point at which it was interrupted) and then polls each subsystem to establish which was the instigator of the interrupt request. Having established the source of the interrupt, the processor then executes an appropriate interrupt service routine before returning to the previously suspended task.

A polled interrupt system is clearly quite simple to operate as far as hardware is concerned. The disadvantage of the technique lies in the time taken to poll the various subsystems until the source of the interrupt is located. In a complex system (where many slave cards are present each regularly generating interrupt requests) an appreciable amount of time can be wasted in the polling process.

14 Introduction to bus-based systems

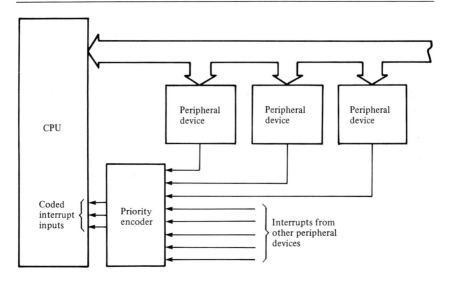

Figure 9 *Multiline interrupt system*

Multiline interrupt systems are individual interrupt request lines connected to a hardware prioritizing arrangement. The encoder (see Figure 9) will indicate to the processor which of the subsystems currently seeking interrupt service has the highest priority. Assuming that interrupts have not been masked, the processor can then suspend its current task and execute the necessary service routine without any further delay.

In a bus-vectored interrupt system, the interrupting subsystem places a 'vector' on the data bus which identifies the source of the interrupt. This data is read by the processor when it responds to the interrupt request and is used to form the memory address of the interrupt service routine. Bus-vectored interrupt systems are thus eminently suited to backplane bus systems and allow for the connection of a large number of interrupting devices (an 8-bit data path will permit the use of up to 255 different interrupting devices, each with its own unique vector).

Desirable characteristics of bus systems

At this point, and before introducing some of the most popular bus standards, it is worthwhile summarising some of the most desirable characteristics of a bus system. These features are not presented in any particular order and their relative importance will, of course, vary according to the application that one has in mind.

In general, a bus system should:

- Be processor and manufacturer independent.
- Allow the use of multiple masters.
- Permit asynchronous operation.
- Employ simple non-multiplexed data transfer protocol.
- Use a simple low-cost backplane.
- Maintain constant impedance of signal tracks within the backplane.
- Permit as high a bus data rate as possible (thus minimizing processing delays).
- Allow as wide an addressing range as possible (both in relation to memory and I/O space).
- Support as wide a range of processors as possible (including 8, 16 and 32-bit types).
- Use a reliable connector (preferably an indirect type).
- Use standard card sizes (e.g. single or double Eurocard).
- Require no 'off bus' connections or links between bus cards.
- Incorporate some means for signalling bus errors.

Common bus systems

We shall now briefly discuss some of the more popular bus systems to have emerged in the last fifteen years. Several of these systems must now be considered obsolete and largely superseded by more modern standards. They have been included in the list so that readers can gain some idea of the development of bus systems generally and, in particular, the emergence of current standards such as STE and VME.

S-100 bus

The S-100 bus was developed by MITS for use in its Altair 8080-based microcomputer system. The system used a number of printed circuit card modules connected into a card frame chassis by means of a double-sided fifty-way direct PCB edge connector. The S-100 bus quickly became the de facto standard for the early 8080-based modular microcomputer systems.

The S-100 bus was later refined so that a number of incompatibility problems were reduced (if not eliminated altogether) and the revised standard is defined under IEEE-696.

The following points are worthy of note:

1 The various supply rails are unregulated and thus on-board regulators are required.
2 There is a risk of cross-talk between some adjacent bus lines on the backplane. For this reason, and to reduce propagation delays in the

backplane, the electrical length of the backplane must be kept reasonably short.
3 Data lines are unidirectional (rather than bidirectional). Thus two sets of eight data lines (data in and data out) are provided. If necessary, these two sets of lines can be combined to provide a bidirectional data bus using an 8-bit bidirectional bus transceiver.
4 The revised S-100 standard (IEEE-696) makes provision for 16-bit processors by extending the address bus from sixteen to twenty-four lines and by ganging the 8-bit data in and data out buses into a 16-bit bidirectional bus. Additional handshaking lines are included in order to permit intermixing of 8- and 16-bit memory cards.
5 The data rate of any signal on the bus should not exceed 6 MHz.
6 The overall structure of the bus is sixteen data lines, sixteen (or twenty-four) address lines, eight status lines, nineteen control lines (five output, six input and eight DMA), eight vectored interrupt lines, twenty utility bus lines and five power lines. Four lines are reserved for future use and three further lines are undefined and thus available for use by individual manufacturers who should clearly specify their logical function. The voltage level of any signal on these lines should not exceed 5 V.

IEEE-488/GPIB bus

While not strictly a backplane bus, the IEEE-488 bus is worthy of mention in the general context of bus systems by virtue of its immense popularity in the field of instrumentaiton. The IEEE-488 bus was originally developed by Hewlett-Packard and is now known simply as the general purpose instrument bus (GPIB). The bus provides a means of interconnecting instruments in automatic test equipment (ATE) configurations where data can be exchanged between a number of participating devices.

The IEEE-488 bus provides for the following types of device:

- Listeners (which receive data from other instruments but are not themselves capable of generating data).
- Talkers (which are only capable of outputing data on to the bus).
- Talkers/listeners (which both generate and receive data).
- Controllers (which manage the flow of data on the bus and provide processing facilities).

While only one talker can be active at any instant, it is possible for several listeners to be receiving data simultaneously.

The IEEE-488 bus uses eight multipurpose bidirectional data lines. These lines are used to transfer data, addresses, commands and status bytes. In addition, five bus management and three handshake lines are provided.

Commands are signalled by taking the attention line (ATN) low. Commands may be directed to individual devices by placing a unique address on the lower five data bus lines. Alternatively, universal commands may be simultaneously directed to all participating devices.

The specified maximum data rate for the bus is normally either 1 or 2 megabyte/s but, in practice, typical data transfer rates are 250 kilobyte/s or less. The maximum terminated bus length is normally assumed to be 20 m.

An important feature of the bus is that it provides for data transfer between services which have widely different response times. In practice, therefore, the slowest listener determines the rate at which data transfer takes place.

A variety of dedicated VLSI GPIB bus interface devices are available. These include Intel's 8291 GPIB listener/talker and 8292 GPIB controller, as well as Motorola's 68488 GPIB adaptor.

The IEC-625 bus is similar to the IEEE-488 bus but employs a different connector (25-way D-type rather than the 24-pin connector originally specified by Hewlett-Packard).

The G-64 bus

The G-64 bus concept was created by GESPAC SA of Geneva in 1979 and it arose from a need to provide industry with an up to date bus standard. The G-64 bus was originally designed for 6809 signals but the specification has been extended to cope with a wider range of processors. Hence the G-64 bus now caters for 8- and 16-bit processors (sixteen individual data lines are provided) and permits asynchronous as well as synchronous data transfer. The G-64 bus is based on the single Eurocard format with a standard module size of 100 mm × 160 mm. The connector specified for use with the G-64 bus is the popular and highly reliable IEC 603–2 (DIN 41612 type b). This connector is an indirect type double-sided 64-way connector with ground connections at each end of the connector (pins 1 and 32 on each side). The G-64 bus employs an unterminated backplane and this limits the speed of data transfer on the bus.

STD bus

The STD bus system was another early contender for the low-cost 8-bit bus market. It provides an 8-bit data path and 16-bit address path. The memory addressing range is 64 K while the I/O space is only 256 bytes (though this has been expanded in a number of implementations). The bus will not directly support multiple masters (unless an off-bus connection is employed). The system is non-Eurocard based (unlike most of its rivals) and it uses a direct edge connector.

As with S-100, various modifications and extensions have been made to the STD standard with a view to extending its life. Unfortunately, these modifications have created a number of problems for end-users relating to compatibility of system cards. The STE bus standard provides a far superior environment for the development of 8-bit systems and many users of the STD bus are converting to this new standard.

Multibus

Multibus was developed by the Intel Corporation and forms the basis for the IEEE-796 bus standard. Multibus permits both 8- and 16-bit data transfers and has a 24-bit address bus. Daisy chained bus priority allows for the presence of typically up to four bus masters. Eight interrupt lines are provided and these may be decoded on a master central processing unit (CPU) card in order to generate the appropriate vectored service routine.

Multibus cards have two connectors: an 86-way edge connector (P1) on which all signal lines are defined and an auxiliary 60-pin connector (P2) for customized functions. The main connector offers regulated power rails of +5 V and +12 V together with eight separate ground connections.

Bus signals are grouped together according to their logical function and placed physically adjacent on the edge connector. Twenty address lines and sixteen data lines are provided and all bus signals are active low (ground true).

Multibus is now defined under IEEE-796 and the successful transition, from 8- to 16-bit operation has gained it a wide measure of support in industry today.

Q-bus and Euro Q-bus

The Q-bus traces its origins from DEC's Unibus system which was prominent in the 1970s and implemented on the popular DEC PDP-11 minicomputer. The system is now used in conjunction with DEC's LSI-11 microcomputer.

Euro Q-bus is based on the Q-bus and employs a single DIN 41612 connector (three rows are used) which provides 22-bit addressing, direct memory access (DMA), and a priority interrupt structure. Euro Q-bus cards are double height (6U) Eurocards derived from DEC LSI-11 modules. Ground connections are made at the ends of the connector (pins 1 and 32 of all three rows) and at four points in row A. In common with most modern bus systems, the bus is fitted with terminating networks.

Versabus

Versabus is an elegant asynchronous bus system which was primarily designed for use with Motorola's 16/32-bit 68000 microprocessor. The bus is not, however, confined to the 68000 device and has considerable capacity for further expansion. Two double edge connectors are used: the basic 16-bit data path being associated with one and the necessary bus lines for a full 32-bit data path and 32-bit addressing being provided on the second connector.

VME bus

The VME bus standard resulted from a joint effort by semiconductor manufacturers Mostek, Motorola, and Signetics. The bus supports very fast data transfer rates (up to 24 MHz) and uses a single 96-way DIN 41612 type c indirect connect to provide a 16-bit data path. A second 96-way indirect connector may be used to provide full 32-bit operation. The basic system incorporates a 24-bit address bus while the extended bus supports 32-bit addresses.

The VME bus has its roots in Versabus and is arguably the highest performing 16/32-bit microprocessor bus system currently available, offering true 'state of the art' performance. The bus is, however, expensive to implement and is thus inappropriate in a number of more basic applications.

STEbus

The STEbus was originally conceived as a Eurocard replacement for the STD bus. Many of the shortcomings of earlier 8-bit bus standards (such as G-64 and STD) have been eliminated. The system was first specified in 1982 and the first STE cards appeared in 1984. Since 1985 the standard has attracted considerable support and many users of earlier 8-bit bus systems have changed to STE. In 1987 the system was defined under the IEEE-1000 specification which further establishes it as an internationally recognized standard.

The STEbus provides for an 8-bit data path and twenty address lines (permitting 1 megabyte of directly addressable memory space). The system provides for an I/O space of 4 kilobytes and up to three potential bus masters may be present within a system.

STE bears more than a passing resemblance to VME and, some enthusiastic users have likened it to an 8-bit implementation of the VME standard. One of the beauties of the STEbus is that, while the system is flexible enough to permit mixing of a wide variety of processors from different manufacturers, it is definitive in terms of bus signals and

Figure 10 Arcom's IBM PC compatible processor running on the rugged STE-bus system (courtesy of Arcom Control Systems Ltd)

protocol. System designers can thus have every confidence that STE bus products obtained from a variety of sources can be interconnected and the whole system will operate as planned.

The STE bus is considered in greater detail in Chapter 3. Furthermore, the majority of applications discussed in this book relate to this standard.

Table 1 Comparison of three popular 8-bit backplane buses

	STD	G-64	STE
Signal lines	56	64	64
Address range (bytes)	64 K*	64 K	1 M
Data path (bits)	8	8	8
I/O space (bytes)	256*	1 K	4 K
Multiple masters	Off bus connection required	Yes	Up to 3
Error checking	None	Yes	Error line
Normal card size (mm)	114 × 165	100 × 160	100 × 160
Connector	Direct edge	DIN 41612 (type b)	DIN 41612 (type c)

*Various expansion schemes have been developed.

2
Introduction to microcomputers

This chapter aims to provide readers with a brief introduction to microprocessors and microcomputer systems. Readers already familiar with microprocessor-based systems need not concern themselves very much with the material presented in this chapter; it is merely provided for those who are unfamiliar with the topic and who require some background information on which to build.

Microprocessors

Microprocessors are sophisticated VLSI integrated circuit devices which are capable of decoding and executing instructions presented to them in binary coded form. Most modern microprocessors perform fundamental operations on groups of either eight, sixteen, or thirty-two binary digits (bits). They are thus divided into 8-bit, 16-bit, and 32-bit types.

An 8-bit microprocessor thus fetches and outputs data in groups of 8-bits (bytes). This data is moved around on a data bus comprising eight separate lines (D0 to D7). The source or destination of the data is determined by a unique address placed on a separate address bus (which, in the case of the vast majority of 8-bit microprocessors, comprises sixteen lines). The microprocessor outputs an address on this bus and then either reads the data present on the data bus during a *read operation* or places data on the data bus during a *write operation*.

A 16-bit microprocessor fetches and outputs data in groups of 16-bits and the data bus comprises sixteen separate lines (D0 to D15). The number of address lines provided on 16-bit microprocessors tends to be either sixteen, twenty or twenty-four.

In the case of a 16-bit address bus, the lines are labelled A0 to A15 and the maximum address range is given by 2^{16} (i.e., 65536 different addresses are possible with decimal addresses ranging from 0 to 65535).

In the case of a 20-bit address bus, the address lines are labelled A0 to A19 and the maximum address range is 2^{20} (i.e., 1048576 individual address locations with decimal addresses ranging from 0 to 1048575). Note that, at each unique address location, 1 byte (8 bits) of information can be stored.

The various control signals required by the system are distributed on a third bus. Particularly noteworthy among these control bus signals are those used for signalling the direction of data movement (i.e., whether a read or write operation is being performed).

Other control bus lines are used for rather more specialized (but none the less important) functions such as providing a means of resetting the system or indicating that an external device requires the attention of the microprocessor. A control bus generally has between five and thirteen individual lines.

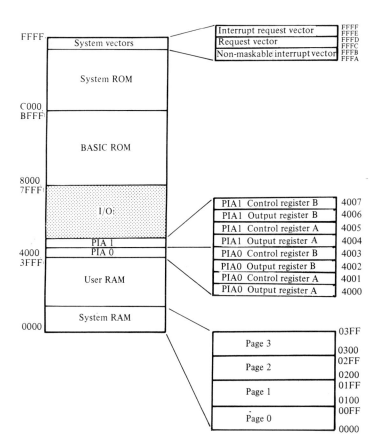

Figure 11 *Typical memory map for a 6502-based microprocessor system*

Memory maps

The address to which the data is to be written or from which it is to be read can either constitute part of the memory of the system (in which case it lies either in RAM or ROM) or can be considered to be associated with input/output (I/O). The allocation of the memory within a system is shown in a memory map. A typical memory map for an 8-bit 6502-based microprocessor system is shown in Figure 11. Note that the memory has been divided into four blocks having the following functions:

1. 16 kilobytes of read-write memory (RAM) for system use and for temporary storage of the user's programs and data.
2. 16 kilobytes reserved for I/O (of which only a small proportion is actually used).
3. 16 kilobytes for permanent storage of a high-level language interpreter (contained within ROM).
4. 16 kilobytes for permanent storage of an operating system (also within ROM).

Readers should note that not all microprocessors require their I/O space to be *memory mapped*. Several microprocessor families employ separate control lines to distinguish between memory and I/O operations. Examples of such devices are the popular 8085 and Z80 devices which distinguish between memory and I/O in the following manner.

The 8085 has a single line (IO/$\overline{\text{M}}$) which is:

1. Taken high when the microprocessor is executing an I/O read or write operation.
2. Taken low when the microprocessor is reading or writing from memory.

The Z80 has two separate lines ($\overline{\text{IORQ}}$ and $\overline{\text{MREQ}}$) which function as follows:

1. The $\overline{\text{IORQ}}$ line is taken low when the microprocessor is performing an I/O read or write.
2. The $\overline{\text{MREQ}}$ line is taken low when the microprocessor is performing a memory read or write.

Internal architecture

The principal internal elements of a microprocessor are as follows:

- Registers for temporary storage of instructions, data and addresses.
- An arithmetic logic unit (ALU) able to perform a variety of arithmetic and logic functions.
- Control logic which accepts and generates internal and external

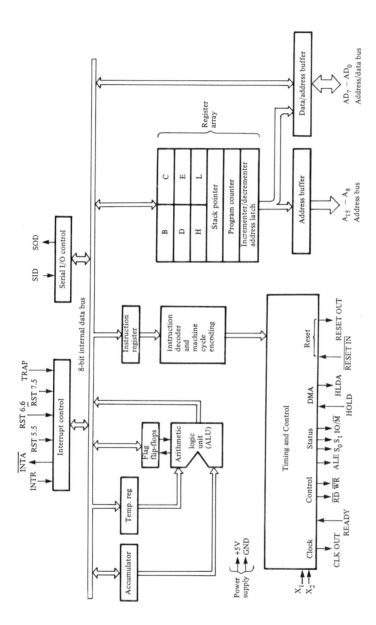

Figure 12 Internal architecture of the 8085 microprocessor

control signals (such as $\overline{\text{READ}}$ and $\overline{\text{WRITE}}$) and provides timing signals which are used to synchronize data transfers.

The internal arrangement (or *architecture*) of a microprocessor tends to vary from one family to another. There are, however, a number of common themes. The major microprocessor families, for example, tend to retain a high degree of upward compatibility both in terms of internal architecture and the software 'instruction set' and this is clearly an important consideration in making a new product attractive to the manufacturer of microprocessor-based equipment.

Figure 12 depicts the internal architecture of a representative microprocessor, the 8085. Despite its apparent complexity, a major portion of this diagram simply represents a collection of registers linked together by means of a high speed internal data bus.

The various internal registers of a microprocessor are simply arrangements of bistable latches into which data (in binary form) can be placed during processing. Some registers are directly accessible to the programmer (i.e., their contents can be modified at will) while others are reserved for the machine's own use. Registers may also be classified as *dedicated* (i.e., they are reserved for a specific purpose such as pointing to a memory location or holding the results of an ALU operation) while others are described as *general purpose*.

The following registers are particularly noteworthy:

Instruction pointer or program counter

The instruction pointer (IP) (or program counter (PC)) is a dedicated register which contains the address of the next instruction byte to be executed. The contents of the register is automatically incremented each time an instruction byte is fetched.

Accumulator

The accumulator (A) functions both as a source and destination register; not only is it the source of one of the data bytes required for an ALU operation but it is also the location in which the result of an ALU operation is placed.

Condition code register, flag register, status register, and program status word

This register contains information on the internal status of the microprocessor and, in particular, signals the result of the last ALU operation. It is important to note that the flag register is simply a collection of bistable latches which are set or reset depending upon the outcome of a previous ALU operation; the state of each bistable can be considered to represent a 'flag'. Commonly available flags are: carry (C), subtract (N), overflow (V) and zero (Z).

Stack pointer

Most microprocessors need to have access to an external area of read/write memory (RAM) which permits temporary storage of data in the form of a stack. The stack operates on a last-in first-out (LIFO) basis; data is 'pushed' onto the stack and later 'pulled' off it. The *stack pointer* (SP) keeps track of the extent of the stack by holding the address of the last used (or next available) stack location. Readers should note that the stack is a dynamic structure which changes in size during processing and that some microprocessors, for example, the 6502, require that the stack is resident in a particular area of memory (page 1 in the case of the 6502). Furthermore, some microprocessors, for example the 6809 and 68000, have two independent stack pointers: a *user stack pointer* (USP) and a *system stack pointer* (SSP) in the case of the 6809 or a *supervisor stack pointer* (SSP) in the case of the 68000.

Instruction register

The instruction register is not directly accessible to the programmer but is used to contain the current instruction byte so that it can be decoded by an arrangement of logic gates known as an *instruction decoder*. The outputs from the instruction decoder are passed to the microprocessor's control logic which, among other things, determines the direction of data transfers and responds to external signals which arrive on the control bus.

Having dealt with the internal features of a typical microprocessor it is worth making some mention of the function of each of the more important external control lines.

Read/write

The read/write (R/$\overline{\text{W}}$) line is taken low when the microprocessor is performing a write operation or high when the microprocessor is performing a read operation. Some microprocessors, for example the Z80, have separate $\overline{\text{READ}}$ (RD) and $\overline{\text{WRITE}}$ (WR) lines.

Interrupt request

The interrupt request ($\overline{\text{IRQ}}$, $\overline{\text{INT}}$) line serves as an input to the microprocessor and is taken low by an external device wishing to signal the fact that it requires attention. Provided the *interrupt flag* is reset (i.e., logic 0) this request will be honoured and the microprocessor will cease normal processing and execute the required *interrupt service routine*.

Non-maskable interrupt

The response to an ordinary interrupt ($\overline{\text{IRQ}}$ and $\overline{\text{INT}}$) is determined by the microprocessor's internal status which may be programmed so that interrupts may be 'masked' (disabled). The method for altering the interrupt status varies with different microprocessor types. The 8085, for example, has software instructions which can be used to set (SIM) or read (RIM) the interrupt mask. The 6502, on the other hand, has an interrupt status flag (I) which can be set or reset to disable or enable interrupts.

There are, however, some situations in which it is desirable that an interrupt should be serviced regardless of what else is going on. Hence a separate *non-maskable interrupt* ($\overline{\text{NMI}}$) line is provided. When this line is taken low, normal program execution is interrupted regardless of the state of the interrupt flag (i.e , regardless of whether interrupts are currently enabled or disabled).

Reset

Reset ($\overline{\text{RES}}$), an active low input to the microprocessor, is used to initialize the system into a known state prior to normal execution of the program. When the reset line is taken low, the program counter (PC) is placed in a defined state (either by loading it with zero or by placing a predefined address 'vector' in it) and interrupts are disabled. Various other internal operations are performed which tend to vary from processor to processor.

Clocks and speed of execution

In order that data flow within a microprocessor is orderly, it is necessary to synchronize all data transfers with its immediate support devices (onboard RAM, ROM and I/O) using a clock signal. This signal may be generated by an external oscillator or equivalent circuitry may be provided inside the microprocessor chip. For accuracy and stability, microprocessor clocks are invariably crystal controlled and typically operate in the range 1 MHz to 8 MHz.

A *clock cycle* (known as a *T-state*) is the fundamental timing interval used by the microprocessor. A *machine cycle* (*M-cycle*) is the smallest indivisible unit of microprocessor activity and usually comprises between three and five T-states. An *instruction cycle* (i.e., that associated with fetching an instruction, decoding and executing it) normally requires between one and five M-cycles.

To give readers some idea of how short this time interval is, suppose a Z80 microprocessor operates with a 4 MHz clock. The fundamental clock cycle (T-state) will then be 250 ns and a machine cycle (M-cycle) will occupy between 0.75 μs and 1.25 μs. A complete instruction cycle will take place in an interval of between 1.25 μs to 6.25 μs, depending upon its complexity. To put this another way, the microprocessor will execute between 160,000 and 800,000 instructions every second.

Microprocessor systems

Even the most sophisticated microprocessor requires the support of a number of other very large scale integration devices. The most common support devices found in almost every microcomputer system are random access memories (RAM), read-only memories (ROM), and input/output (I/O) devices.

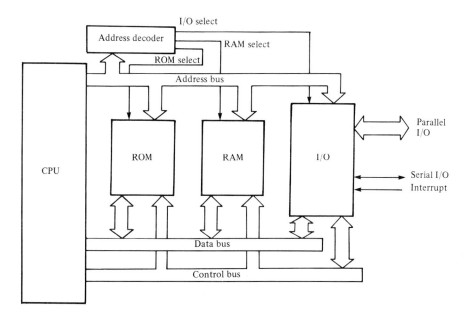

Figure 13 *Basic configuration of a microprocessor system*

The basic configuration of a microprocessor system showing the internal elements (microprocessor, ROM, RAM and I/O) is shown in Figure 13. The elements are linked by means of three buses used for data, addresses and control signals. In the simple system of Figure 13, the only device capable of generating address information is the microprocessor and hence the address bus is unidirectional. The data bus, on the other hand, must clearly be bidirectional since data must be transferred to and from the microprocessor and its various support devices. In more complex systems, it will often be desirable for sophisticated support devices to be able to drive the address bus in order to facilitate fast direct memory access (DMA) data transfers. In such systems, the microprocessor must be capable of placing its address bus drivers in a tristate condition (thus effectively isolating itself from the address bus) so that other devices can drive the bus without risk of conflict.

Bus connected support devices (such as ROM, RAM, etc.) have tristate data bus drivers and are brought into use by asserting their enable ($\overline{EN}$) or chip select ($\overline{CS}$) lines. These conventionally active-low signal lines are driven by address decoding logic with inputs driven from the address bus. Hence, where the processor is reading and writing to RAM, for example, the address decoding logic will ensure that only the RAM is selected and the internal buffers in the ROM and I/O chips are kept in the tristate output condition. This topic is discussed in greater detail later in this chapter.

Memories

While some specialized microprocessors (of the type used in simple low cost stand-alone control systems) incorporate limited amounts of internal memory for data and program storage, most general purpose microprocessors require several types of external memory device. These tend to fall into two distinct categories: *read/write* or *read/only*.

Read/write memory is simply memory which can be read from and written to. In other words, the contents of the memory can be modified at will. Read-only memory, on the other hand, can only be read from: an attempt to write data to such a memory will have no effect on its contents.

Microprocessor systems use semiconductor memories for compact and high-speed storage and retrieval of data. Semiconductor read-only memories (ROM) are available in various forms depending upon the particular application concerned. Semiconductor read/write memory is usually implemented by so-called *random access* memories (RAM). The term random access simply indicates that one can access data stored anywhere within the memory with equal ease.

Read-only memory (ROM)

Microprocessors require non-volatile storage for their control programs and, where appropriate, operating systems and high-level language interpreters. This is an ideal application for a ROM. Where the control program is to be updated or the operating system modified, the ROM can be removed and replaced with a later version.

Mask programmed read-only memories
When large scale production of a microprocessor-based system is envisaged (as is the case with a personal computer, for example), the most cost-effective method of implementing read-only memory is with the aid of a mask programmed device. Such devices are programmed during manufacture, the mask used determining the actual data stored within the chip. The programming information is supplied to the semiconductor manufacturer by the manufacturer of the microprocessor system. Since the process is only cost effective for quantities in the tens of thousands, it should be obvious that the manufacturer of the microprocessor system needs to be very confident that the stored data and program is completely free from errors and will not require further modification.

Fusible-link programmable read-only memories
Fusible-link programmable read-only memories (PROM) are cost effective for medium scale production and are programmed by the equipment manufacturer rather than the semiconductor manufacturer. The PROM consists of an array of nichrome or polysilicon fuses. These fuse links may be 'blown' by applying a current pulse of sufficient magnitude to rupture the link. Programming takes a considerable time but the equipment required is simple and relatively inexpensive. Often, early variants of microprocessor systems are supplied with PROM devices which are later replaced with mask ROM devices as soon as teething troubles and bugs have been eliminated and large scale production commences.

Erasable programmable read-only memories
The programming of mask-programmed and fusible-link memories is irreversible. Once programmed, devices cannot be 'wiped clean' in preparation for fresh programming. The erasable programmable read-only memory (EPROM) is, however, capable of erasure allowing the device to be programmed and reprogrammed many times over.

EPROM devices are fitted with a window which allows light to fall upon the memory cell matrix. When this area is exposed to strong ultraviolet (UV) light over a period of several minutes, the stored data is erased. The EPROM may then be reprogrammed using a low cost device which supplies pulses of current to establish the state of individual memory cells. This process can take up to several minutes though some EPROM programmers can program several devices at once.

EPROMs are ideal for small scale production and software development. They are, however, relatively expensive and thus are not generally suitable for large scale production. The data in Table 2 refers to some of the most popular EPROM devices.

Table 2

Type	Size (bits)	Organization	Package
2516 / 2716	16384	2 K words × 8 bits	24-pin DIL
2532 / 2732	32768	4 K words × 8 bits	24-pin DIL
2564 / 2764	65536	8 K words × 8 bits	28-pin DIL
27128	131072	16 K words × 8 bits	28-pin DIL
27256	262144	32 K words × 8 bits	28-pin DIL
27512	524288	64 K words × 8 bits	28-pin DIL

Electrically alterable/electrically erasable read-only memories

The electrically alterable read-only memory (EAROM) and electrically erasable read-only memory (E^2ROM) are relative newcomers which, unlike the EPROM, can be erased electrically thus avoiding the need for removal from the equipment to which they are fitted. Unfortunately, EAROM and E^2ROM devices are rather expensive and, owing to the availability of low-power CMOS RAM devices which may be *battery backed*, they have not as yet made much impact on the industry as a whole.

Readers may be forgiven for thinking that EPROM and EAROM devices are really forms of read/write memory. In a sense they are but a distinction must be made between these memories and 'true' read/write memories (for example semiconductor RAM) in which individual bytes can be changed at will. Another important point is the time and ease with which a device may be reprogrammed with data. A single byte of data can be accessed from a semiconductor RAM in a typical time interval of 150 ns. An entire 8 K RAM can have its data changed in a time interval of 8192 × 150 ns (plus an additional overhead for the processor). Programming times are thus very much faster than those associated with equivalent capacity EPROM devices which may require several minutes for programming not to mention the time taken for removal and erasure under UV-light.

Random access memory (RAM)

Microprocessors also require access to read/write memory in the form of semiconductor RAM. A portion of this memory is used by the operating system for the storage of 'system variables'. The operating system and control program also require access to RAM for temporary storage of data in the form of one or more *stacks*. A further area must be made available for users to store their own programs and data. In addition, where a raster scanned display is used, an area of RAM is usually devoted to a *screen memory*. With modern systems, this reserved area of memory is invariably *bit mapped* (i.e., each bit of *screen RAM* corresponds to a particular pixel).

Bipolar random access memories
Bipolar RAM devices employ a conventional bistable arrangement based upon bipolar transistors. These memories require considerable power and are only available with relatively small capacities. They are, however, extremely fast and are thus used in high speed applications and to implement buffers between fast processing devices and conventional memories which operate at much slower speeds.

NMOS static memories
NMOS static memories are also based upon a bistable cell configuration. NMOS devices require considerably less power than their bistable counterparts thus permitting much greater densities to be achieved.

CMOS static memories
CMOS static memories use cell configurations which are similar to those used by their NMOS counterparts. CMOS devices, however, consume negligible power when they are not operating so they are ideal for equipment which has to be operated from a battery power supply.

NMOS dynamic memories
Unlike the previously mentioned memory types, NMOS dynamic memories are based upon charge storage within a basic memory cell rather than the action of a bistable element. Since the value of capacitance that can be produced within a cell is extremely small, the charge stored leaks away rapidly and thus dynamic memories require regular 'refreshing'. The process of refreshing a dynamic memory involves periodically reading the data stored and then writing it back to the memory. This process may be undertaken by the microprocessor as part of its normal activities or may be consigned to a dedicated dynamic memory controller chip.

The data in Table 3 refers to a range of the most popular semiconductor devices:

Table 3

Type	Technology	Size (bits)	Organization	Package
2114	NMOS static	4096	1 K words × 4 bits	18-pin DIL
2128	NMOS static	16384	2 K words × 8 bits	24-pin DIL
4116	NMOS dynamic	16384	16 K words × 1 bit	16-pin DIL
4118	NMOS static	8192	1 K words × 8 bits	24-pin DIL
4164	NMOS dynamic	65536	64 K words × 1 bit	16-pin DIL
4256	NMOS dynamic	262144	256 K words × 1 bit	16-pin DIL
4464	CMOS static	65536	8 K words × 8 bits	28-pin DIL
4864	NMOS dynamic	65536	64 K words × 1 bit	16-pin DIL
6116	CMOS static	16384	2 K words × 8 bits	24-pin DIL
6264	CMOS static	65536	8 K words × 8 bits	28-pin DIL
41256	NMOS dynamic	262144	256 K words × 1 bit	16-pin DIL
43256	CMOS static	262144	32 K words × 8 bits	28-pin DIL
50464	NMOS dynamic	262144	64 K words × 4 bits	18-pin DIL

Address decoding

Each location in semiconductor ROM and RAM has its own unique address. At each address a byte (comprising 8 bits) is stored. Each ROM and RAM (or bank of RAM devices) accounts for a particular block of memory, its size depending upon the capacity of the ROM or RAM in question. As an example, a simple Z80-based system may have two 16 K ROM and two 16 K blocks of RAM (each comprising eight 16 K × 1 bit RAM chips) to make up a total memory space of 64 kilobytes. These devices may have addresses assigned as given in Table 4.

Table 4

Device	Capacity (kilobytes)	Address range Hexadecimal	Decimal
ROM 0	16	0000–3FFF	0–16383
ROM 1	16	4000–7FFF	16384–32767
RAM 0	16	8000–BFFF	32768–49151
RAM 1	16	C000–FFFF	49152–65535

Each RAM device will have fourteen address input lines (A0 to A13) and one chip select ($\overline{CS}$) line. The chip select line is invariably active-low hence, to enable the output of a selected RAM or ROM onto the bus, it will be necessary to ensure that this line is taken low. In addition, all of the semiconductor memory devices will be connected to the read/write line (or a derivative in the form of a memory read/write line).

The address lines of each RAM are connected (together with those of the ROM) to their respective lines on the address bus. Hence, unless something is done to prevent a conflict, we would find that all four blocks of semiconductor memory were being written to or read from at the same time. A method of decoding the two most significant address lines (A13 and A14) so that one (and only one) of the chip select ($\overline{CS}$) lines is enabled at any time. A suitable decoding scheme is shown in Table 5.

Table 5

| Address lines | | Device enabled |
A15	A14	
0	0	ROM 0
0	1	ROM 1
1	0	RAM 0
1	1	RAM 1

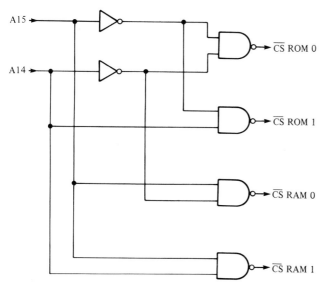

Figure 14 *Simple address decoder based on standard logic gates*

The problem of decoding the two most significant address lines may be easily resolved using some fairly simple logic based on arrangements of standard logic gates (Figure 14) or could form part of a larger programmed logic array (PLA). Alternatively, a decoder or demultiplexer may be used as shown in Figure 15. Note that the 74LS139 also has an active-low enable input ($\overline{EN}$). This can be used to disable the chip so that the entire semiconductor memory is simultaneously disabled. Such a

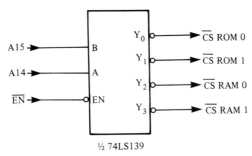

Figure 15 *Address decoder based on a 74LS139 demultiplexer*

provision may prove useful in situations where, for example, it is necessary to *bank* several ROM, RAM or other memory mapped support devices into a common address space.

Input/output devices (I/O)

Clearly, to fulfil any useful function, the microprocessor needs to have links with the outside world which are invariably supplied by means of one or more VLSI devices. These I/O devices fall into two general categories: *parallel* (a byte is transferred at a time) or *serial* (1 bit is transferred after another along a single line). Both types may be configured under software control and are thus said to be programmable.

Memory mapped I/O versus port I/O

As mentioned earlier, a basic distinction exists between systems which treat I/O devices as one or more memory addresses and those which assign specific port addresses to each I/O device. In either case, data is output by simply writing data to an appropriate memory or port address. Conversely, data is input by simply reading it from the appropriate memory or port address. Using memory mapped I/O, the CPU treats I/O operations in exactly the same way as operations performed on memory. A proportion of the memory space is then reserved for I/O (it cannot, at the same time, be assigned to RAM or ROM!). Port I/O, on the other hand, reserves a set of port addresses which are quite independent of the conventional memory space. Port addresses are distinguished from memory addresses by means of the $\overline{\text{IORQ}}$ and $\overline{\text{MREQ}}$ control bus signals and instructions are provided for port input and output.

Parallel versus serial I/O

Another important distinction has to be made between parallel and serial I/O. In the former case, a byte of data is transferred at a time (thus necessitating an 8-bit buffer to latch) whereas in the latter case, data must be transferred a bit at a time. Parallel I/O invariably implemented by one or more programmable devices which can be configured in a variety of different modes depending upon individual requirements.

Since the data present on a microprocessor bus exists primarily in parallel form (it is byte wide) serial I/O is a little more complex. Serial input requires a means of conversion of the serial input data to parallel data in a form which can be presented to the bus. Serial output, on the other hand, requires a means of conversion of the parallel data present on the bus into serial output data. In the first case, conversion can be performed with a serial input parallel output (SIPO) shift register while in the second case a parallel input serial output (PISO) shift register is required. Again, the most effective method of implementing such an interface is with the aid of a dedicated programmable VLSI device.

I/O control methods

There are three basic methods of controlling I/O operations. These vary in complexity and flexibility and each has its advantages and disadvantages. The simplest method, and that which is perhaps the most obvious, is that of allowing the microprocessor to control all I/O operations in a strict sequence defined by its control program. This method, known as *programmed I/O* or *polled I/O*, ensures that the processor has full control of the situation but is inflexible and rather slow. Essentially, the processor periodically polls each peripheral (via its respective I/O device) to determine whether it requires service. If a service request is detected, the processor will execute the appropriate service routine. Once a service request has been honoured, any requests for service generated by other peripheral devices will be ignored; they must simply wait until the processor becomes available to service their requests.

A better method, but somewhat more complex, is that of allowing peripheral devices to 'interrupt' the operation of the processor and, depending upon its internal status, the processor will suspend its current task before executing the appropriate service routine. The processor will use the stack to preserve any important parameters as well as the address of the instruction within the main program which should be returned to once the interrupt service routine has been completed.

Interrupts may be *prioritized* in hardware such that the most important peripheral device receives attention first. A particular example might be an emergency stop button fitted to a machine; we would almost certainly want the system to respond to a signal from this source regardless of what else was happening at the time.

Parallel I/O devices

Having attended to the basic concepts of microprocessor system inputs/outputs, we shall turn our attention to some common programmable I/O devices. Parallel I/O devices enjoy a variety of names depending upon their manufacturer. Despite this, parallel I/O devices are remarkably similar in internal architecture and operation with only a few subtle differences distinguishing one device from the next. The following types are commonly encountered:

6520	Peripheral interface adaptor (PIA).
6521	Peripheral interface adaptor (PIA) – similar to the 6520.
6522	Versatile interface adaptor (VIA).
6820	Peripheral interface adaptor (PIA) – equivalent to the 6520.
6821	Peripheral interface adaptor (PIA) – equivalent to the 6521.
8255	Programmable parallel interface (PPI).
Z80-PIO	Programmable input/output (PIO).

As their names imply, programmable parallel I/O devices can normally be configured (under software control) in one of several modes:
1. All eight lines configured as inputs.
2. All eight lines configured as outputs.
3. Lines individually configured as inputs or outputs.

In addition, extra lines are usually provided for *handshaking*. This is the aptly named process by which control signals are exchanged between the microcomputer and peripheral hardware.

The nomenclature used to describe port lines and their function tends to vary from chip to chip but again there is a reasonable degree of commonality. The following applies to most of the devices listed previously:

PA0 to PA7	Port A I/O lines: 0 corresponds to the least significant bit (LSB) while 7 corresponds to the most significant bit (MSB).
CA1 to CA2	Handshaking lines for Port A: CA1 is an interrupt input while CA2 can be used as both an interrupt input and peripheral control output.
PB0 to PB7	Port B I/O lines: 0 corresponds to the least significant bit (LSB) while 7 corresponds to the most significant bit (MSB).
CB1 to CB2	Handshaking lines for Port B: CB1 is an interrupt input while CB2 can be used as both an interrupt input and peripheral control output.

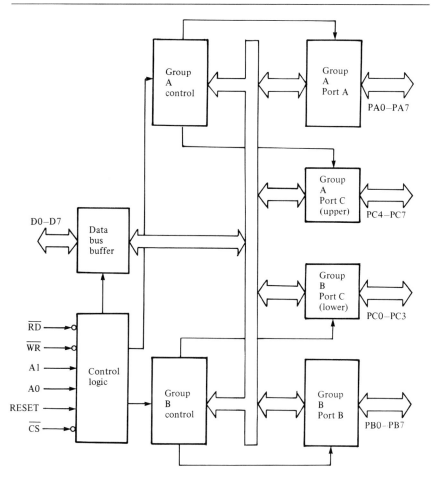

Figure 16 *Simplified internal arrangement of a programmable parallel I/O device*

Figure 16 shows the simplified internal arrangement of a typical programmable parallel I/O device. The two 8-bit I/O ports are quite independent and thus separate data, direction, and control registers are provided. Each of these registers is mapped to a different I/O location or port address.

The electrical characteristics of an I/O port tend to vary from chip to chip. However, signals are invariably TTL compatible. Several programmable parallel I/O devices have port output lines (usually the B group) which are able to source sufficient current to permit the direct connection of the base of a conventional or Darlington type transistor. This device can then be used as a relay or lamp driver. Alternatively, high-voltage open-collector octal drivers may be connected directly to the port output lines.

Serial I/O devices

Serial data may be transferred in either synchronous or asynchronous mode. In the former case, all transfers are carried out in accordance with a common clock signal (the clock must be available at both ends of the transmission path). Asynchronous operation involves transmission of data in *packets*; each packet containing the necessary information required to decode the data which it contains. Clearly this technique is more complex but it has the considerable advantage that a separate clock signal is not required.

As with programmable parallel I/O devices, a variety of different names are used to describe programmable serial I/O devices. The most commonly encountered are listed below:

6850	Asynchronous communications interface adaptor (ACIA).
6852	Synchronous serial data adaptor (SSDA).
8251	Universal synchronous/asynchronous receiver/transmitter (USART).
8256	Universal asynchronous receiver/transmitter (UART).
Z80-DART	Dual asynchronous receiver/transmitter (DART).

As with programmable parallel I/O devices, there is a degree of commonality in the internal architecture of serial I/O devices. Figure 17 shows the typical internal arrangement of a serial I/O device which can be used to implement two separate data communication channels.

Common signals associated with serial I/O devices are listed below:

D0 to D7	Data input/output lines connected directly to the microprocessor bus.
RXD	Received data (incoming serial data).
TXD	Transmitted data (outgoing serial data).
$\overline{\text{CTS}}$	Clear to send. This signal (invariably active-low) is taken low by the peripheral when it is ready to accept data from the microprocessor system.
$\overline{\text{RTS}}$	Request to send. This signal (invariably active-low) is taken low by the microprocessor system when it is about to send data to the peripheral.

The signals generated and received by programmable serial I/O devices are invariably TTL compatible. It should be noted that, in general, such signals are unsuitable for anything other than the shortest of transmission paths (for example, between a keyboard and a computer system enclosure). Serial data transmission over any appreciable distance invariably requires additional buffers and level shifters between the serial I/O device and its associated transmission path.

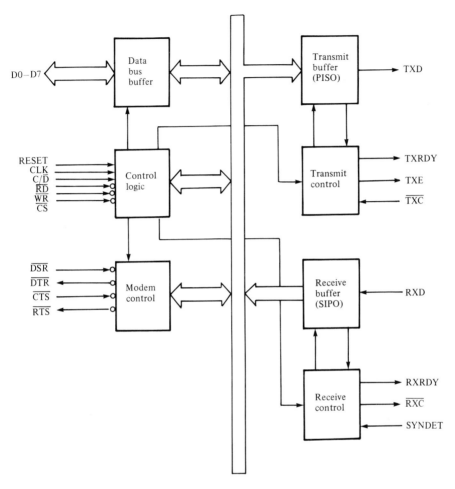

Figure 17 Simplified internal arrangement of a programmable serial I/O device

Operating systems

In general, an operating system is simply a collection of system programs which allow the user to run applications software without having to produce the hardware-specific routines required for such mundane tasks as keyboard and disk I/O. The operating system thus frees applications programmers from the need to be aware of the hardware configuration and present them with what is, in effect, a 'virtual machine' whose characteristics are more tractable than the underlying 'physical machine'.

One obvious advantage of the virtual machine concept is that, provided a form of the operating system is available for a range of machines, a high degree of software portability is ensured.

As a minimum, an operating system can normally be expected to perform the following tasks:
- Accept keyboard entry of commands.
- Create, copy and delete disk files.
- Load programs and data from disk into RAM.
- Save programs and data from RAM to disk.
- Maintain some form of directory (catalogue) of disk files.
- Provide user prompts and appropriate status and error messages.

Operating systems may either exist as firmware in ROM or may be loaded from disk into RAM when the system is initialized (booted) on power-up.

System software

Various items of utility software are required by the software developer. These may include editors, assemblers, interpreters, compilers, linkers, loaders and debuggers.

Editors
Editors allow the user to create and modify text files. Two basic types of editor exist: line editors which only permit operation on a single line at a time and full-screen editors which permit editing over the entire screen area. The action of an editor is similar to that of a simple ASCII word processor. Most assemblers and high level language interpreters and compilers will accept pure ASCII text input, thus an editor can be used to create source text for programs written in a variety of different languages.

Assemblers
An assembler generates machine code (object code) from assembly language source text (source code) presented in mnemonic form. The assembler normally needs to read the source text twice in order to accomplish this task.

On the first pass, the assembler generates a symbol table which is stored in RAM. This table is used to generate absolute addresses from labels (symbolic addresses) used within the source text. On the second pass, the assembler generates the machine code (object code) for each instruction.

Assembler directives can be included within the source text. These pseudomnemonics are not translated into object code but are recognized by the assembler during the assembly process and their purpose is to allow the user to modify, in some way, the object code produced.

Interpreters

Interpreters translate a high level language program (normally presented in the form of an ASCII text file) into machine executable code. An important feature of an interpreter is that it acts on one statement of source code at a time. The statement (often a single line of code) is read, checked for syntax, translated into machine code, and then executed immediately before the next program statement is fetched. The interpreter must, therefore, be permanently resident (it is used whenever the program is run) and execution is often painfully slow by comparison with an equivalent compiled program. The advantage of using an interpreter is that flaws in the source code can be discovered immediately. A statement containing an obvious error will not be accepted by the interpreter, instead, an appropriate error message will be generated, and the user prompted for a correction.

Compilers

Compilers also generate machine code programs from source text written in high level languages. Unlike interpreters, the process of compilation is carried out as a separate operation and the entire program is converted into machine code before execution. The resulting program can then be saved, loaded and run without the compiler being present.

Compiled programs are very much faster than their interpreted counterparts but have the disadvantage that the entire process of compilation must be repeated whenever a change has to be made to the code. This process is generally rather tedious.

Linkers and loaders

After compilation, a program generally needs to be linked with the necessary I/O or mathematical routines contained in a run-time library. The necessary routines are then simply added and linked to the machine code. This process is achieved with the aid of a linker.

Finally, during compilation and assembly, symbolic labels are normally used to represent addresses. Programs can thus be made relocatable (i.e., they can be loaded anywhere in unreserved RAM). The final stage of the process involves replacing all symbolic addresses with absolute addresses to be used for final execution. This is the function of a loader.

Debuggers

Unfortunately, one seldom finds that the code one has just produced runs without a hitch. Debuggers provide a means of testing programs interactively during run-time. Common features of debuggers include:

- Displaying the contents of a block of memory in binary, hexadecimal, or ASCII.

- Filling a block of memory with a given data value.
- Disassembling a block of memory.
- Moving a block of memory from one location to another.
- Examining and modifying the microprocessor registers.
- Reading a disk file into memory.
- Showing the memory layout of a disk file read.
- Writing the contents of a block of memory to disk.
- Loading a program ready for execution.
- Executing a program with optional breakpoints.
- Tracing or single-stepping a program.
- Setting up a file control block and command tail.
- Performing hexadecimal arithmetic.

Programming languages

Microcomputer systems may generally be programmed in a variety of languages and the choice of a particular language will usually depend upon a number of factors including speed of execution, memory space requirement and ease of use.

Various attributes are ascribed to programming languages including *high level* or *low level, structured* and *unstructured*. Each statement of a high level language program has an easily recognized function and thus the code is readable and (usually) easily understood. The function of individual statements of a low level language, on the other hand, may not be very clear and the program will have to be extensively commented if its action is to be understood.

Some languages lend themselves well to structured coding, others do not. The fact that a language permits the use of structures does not, however, imply that all programs written using the language will be structured. Structure is rather more a characteristic of an individual programmer's work; even the most elegant of programming languages can be used to produce relatively unstructured (i.e., messy) code.

The thee basic rules of structured programming are simply that:

1 Overall program flow should be sequential; repeated jumps backwards and forwards in the code should be avoided.
2 Sections of code that are executed repeatedly should only be written out once.
3 All program flow should be explicit and should use appropriate logical constraints (for example, IF . . . THEN, DO . . . UNTIL and WHILE . . . WEND).

A structured approach to programming can easily be developed and simply requires forethought and a little self-discipline on the part of the programmer. The adage 'think first – code last' is particularly appropriate; structured programs are rarely produced by simply sitting at a keyboard and entering a code.

The benefits of adopting a structured approach to program writing are summarized below:

- Code is easier to test and debug (modules may be individually tested before assembling them into a larger program).
- Code is easier to maintain (modules can be easily modified or extended).
- Code modules can be readily transported from one program to another, thus saving time and effort.

Assembly language

The most closely related language to that of the machine itself is assembly language. Assembly language programs use symbolic addresses (instead of actual memory locations) and mnemonic operational codes (opcodes). The assembly language program is translated into executable machine code by means of an assembler.

The principal disadvantage of assembly language is that programs are not readily transportable from one processor family to another. Furthermore, to be adept with assembly language programs the programmer must have an intimate knowledge of the hardware configuration of a system.

The advantage of assembly language programming is that code is extremely efficient, i.e., it is both fast in execution and very compact. Assembly language programs do not need the services of a compiler or interpreter and thus a minimum of additional software (an assembler and debugger) is required in order to produce a functional program. Indeed, in the case of 8-bit processors, short lengths of code may even be 'hand assembled', i.e., the machine code corresponding to a particular assembly language instruction is found by reference to the published instruction set and then simply entered directly into memory using a hexadecimal loader.

BASIC

Beginners all purpose symbolic instruction code (BASIC) was developed at Dartmouth College by John Kemeny and Thomas Kurtz. The principal aim of its creators was to produce a language that the non-programmer would find both acceptable and usable. BASIC statements therefore tend to use simple English words rather than abstract symbols.

BASIC is undoubtedly today's most popular language for a wide range of simple applications and the language is available in a number of forms (including both interpreted and compiled). In many low-cost machines the BASIC interpreter is resident in ROM (and thus instantly available). In other machines BASIC interpreters or compilers are loaded from disk or tape.

BASIC is an extremely easy language to learn but most implementations do little to encourage the user to adopt a structured approach to programming. More powerful BASICs do, however, include such useful features as procedures and user defined functions and the language is still very popular in industrial applications despite the availability of several arguably superior languages.

C

The C programming language was developed at the Bell Laboratories in 1972 by Dennis Ritchie. C is based on a language called B, which was itself a development of BCPL (basic combined programming language).

C uses a relatively small amount of processor-dependent code and is thus highly flexible and portable. This has contributed greatly to the popularity of the language among systems and applications programmers alike. The language favours structured programming and makes an excellent (and highly portable) alternative to the use of assembly language.

C is not particularly easy to learn (particularly if one has already acquired more than a passing acquaintance with relatively unstructured languages such as BASIC). Furthermore, the actions of many common C functions (which should perhaps be better described as procedures) are difficult to comprehend from a mere cursory inspection of the code. That said, C is undoubtedly an exceptionally powerful and highly efficient language.

Forth

Forth was invented by Charles Moore as a means of controlling an astronomical telescope at the Kitt Peak Observatory. It is fair to say that no other high level language is as comfortable or as versatile in real-time control applications as Forth. The breadth and scope of its applications are enormous, from controlling a washing machine to managing a canning plant.

Forth makes extensive use of a resident dictionary and parameter stack. The user is able to define new words and add these to the dictionary. Forth is thus extensible, and this feature makes the language extremely attractive because it allows users to develop their own supplementary dictionary relevant to a particular application.

Forth should be seen as an excellent compromise between BASIC at one extreme and assembly language at the other. Forth, like BASIC, is highly portable and widely available. The code produced is compact and execution speeds can approach those of programs written in assembly language. Despite Forth's obvious suitability for customized control applications, it is disappointing to note that the language is still not gaining the popularity that it deserves.

3
STEbus hardware

The level of performance which may be expected from any microprocessor-based control system is determined, to a large extent, by the means used to effect communication (data transfer) between the various components present within the system. The bus structure (and its associated protocols) is thus crucial to the successful and efficient operation of the system as a whole.

The STEbus can be described as a modern, general purpose, high performance 8-bit backplane bus. STE is as equally at home in stand alone configurations as it is in more sophisticated multiple-bus architectures where it can operate as a secondary bus or high speed I/O channel. Having provided readers with a general introduction to bus-based systems and microcomputers, this chapter deals with the STEbus in greater depth and looks at STEbus hardware in particular.

Functional elements

The STEbus structure is based on the familiar master-slave concept in which a master, having gained control of the bus, may address and command slaves. Masters and slaves communicate with one another by the use of asynchronous interlocked handshake protocol. This technique permits the use of bus devices which operate at widely varying speeds.

The functional elements within an STEbus system fall into the following general categories:

- System controller
- Arbiter
- Master
- Slave

We shall briefly examine the role of each of these basic elements in their relative order of importance within a system as a whole.

System controller

The IEEE-1000 bus specification allows for the presence of only one system controller within any STEbus system. This device is solely responsible for the provision of the following backplane signals:

1. A general purpose clock signal, SYSCLK
2. A power-on reset signal, SYSRST*
3. A transfer error signal, TRFERR*

Note that the functions of a system controller can be combined with those of a master on a single bus card.

Arbiter

The arbiter provides bus grant signals in response to requests from potential bus masters. Only one arbiter can be present within an STE system. As with the system controller, the function of an arbiter may be combined with those of a master on a single bus card.

Master

A master is an element which is capable of controlling the transfer of data on a bus. Naturally, a master must contain a microprocessor or equivalent logic capable of supporting the protocols employed.

All potential bus masters must request allocation of the bus from the arbiter whenever the bus is not in use by another master. A *default master* is necessarily combined with the arbiter on the same board and has the lowest priority for bus allocation. Note that there can only be one default master within an STEbus system.

All other potential masters must request allocation of bus control from the arbiter. Such a request is made by the master asserting one of the two bus request (BUSRQ*) lines. In the case of a master which is on the same board as the arbiter but is not configured as a default master, the request is made by asserting a third bus request line which is private to the board concerned.

Masters may retain control of the bus for a period of time constrained only by the specific requirements of the system. Two basic modes of operation may be used by masters:

1. Release-when-done, in which the master retains control until all required transfers have been accomplished.
2. Release-on-request, where the master retains control of the bus for an

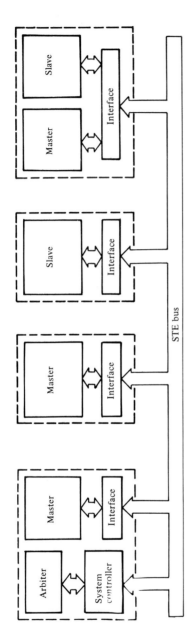

Figure 18 *Various STE bus system configurations*

indefinite period, relinquishing control when it determines that another master requires use of the bus.

Note that a default master operates in release-on-request mode.

Slaves

Boards which are capable of being controlled by the STEbus are designated slaves. As with all bus systems, slaves decode the address lines and act upon the commands generated by bus masters. A slave may also be combined with other functional elements on a board (for example, a single board may contain a master together with read/write memory accessible by other masters in the system).

Various system configurations are shown in Figure 18.

Signal lines

The signal lines used in the STEbus specification are divided into the following five functional groups:

1. Information lines (comprising lines used for data, addresses and commands).
2. Synchronization lines.
3. Attention request lines.
4. Bus allocation lines.
5. Utility lines.

We shall now briefly examine the function of the lines within each of the above groups.

Address lines

The STEbus supports separate address space for memory and I/O; memory transactions reference a 1 megabyte physical address space whilst I/O operations employ a maximum of 4096 (4 K) physical address locations. In either case, the twenty address lines provided by the STE bus are used to specify addresses. In addition, address lines are also used

Table 6

Operation	Address lines used	Lines valid
Memory read or write	20	A19 to A0
I/O read or write	12	A11 to A0
Vector fetch	3	A2 to A0

during a 'vector fetch' response to an attention request in which case they specify the level of the request being acknowledged. As usual, A19 is the most significant address bit while A0 is least significant. The functions of these twenty lines are summarized in Table 6.

Data lines

Eight bidirectional data lines are used to transfer information between masters and slaves. D7 is the most significant bit while D0 is the least-significant.

Command lines

The three command modifier lines (CM2, CM1, and CM0) are used to indicate the type of the current data transfer according to the Table 7.

Table 7

Bus cycle	CM2	CM1	CM0
Reserved	0	0	0
Reserved	0	0	1
Reserved	0	1	0
Vector-fetch	0	1	1
I/O write	1	0	0
I/O read	1	0	1
Memory write	1	1	0
Memory read	1	1	1

Synchronization lines

Four of the STEbus lines are used for synchronization. Two of these are used to indicate that valid addresses or data are present on the respective buses. These lines are known as address strobe (ADRSTB*) and data strobe (DATSTB*). The data transfer acknowledge (DATACK*) line is used to indicate that data has been placed on or accepted from the data bus. The transfer error line (TRFERR*) is asserted when an error is detected during a current data transfer operation.

Attention request lines

The eight attention request lines of the STEbus are used to indicate such user-specific events as interrupt requests and direct memory access requests as well as notification of conditions which exist at either board or system level, for example, power failure.

The attention request lines thus provide a means whereby any board present in a system can request the attention of any other board present. Any board may be connected to any one of the eight attention request lines. Furthermore, several boards may be connected to the same attention request line thus allowing for the broadcast of events to several boards within a system. Readers should note that ATNRQ0* has the highest priority while ATNRQ7* has the lowest priority.

Bus request lines

The STEbus provides two bus request lines (BUSRQ0* and BUSRQ1*) which are used by potential masters to indicate that they wish to control the bus. BUSRQ0* has the highest priority of the two lines.

Bus acknowledge lines

The STEbus provides two bus acknowledge lines (BUSAK0* and BUSAK1*). These are used by the arbiter to indicate to a master requesting bus allocation that it may take control of the bus. BUSAK1* indicates a bus grant to the master requesting via BUSRQ1* while BUSAK0* indicates a grant to the master requesting via BUSRQ0*.

Utility lines

The STEbus system has two further signal lines (SYSCLK and SYSRST*) which are independent of the protocol of other signals. The system clock (SYSCLK) is a 16 MHz square wave generated by the system controller. The clock signal is specified by the timing diagram shown in Figure 19. The system reset (SYSRST*) line is used to place the system in a known initial state. Accesses to the STEbus are inhibited during the period for which SYSRST* is active.

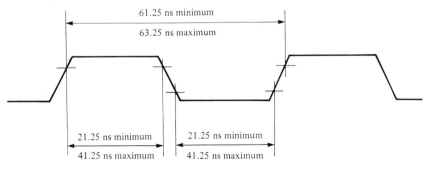

Figure 19 *Timing diagram for the STE clock signal*

The system controller is required to activate the system reset line for not less than 200 ms and not more than 500 ms during power-up. The system controller is also required to assert the system reset signal whenever the system supply falls below its minimum specified tolerance. A reset signal asserted in this manner should remain present for the entire period during which the supply is outside its specified tolerance.

Arbitration

On power-up or immediately following a system reset, the arbiter has control of the bus. Thereafter, all potential masters (except default masters) must request and receive an allocation grant from the arbiter prior to taking control of the bus. The arbiter may employ any algorithm implementable within the constraints of the control transfer protocol to determine which of the bus request levels will be honoured when a current master releases the bus. The algorithm does not form part of the IEEE-1000 specification. Any algorithm implementable within the constraints of the control transfer protocol is acceptable. Note, however, that the BUSRQ0*/BUSAK0* pair conventionally has priority over the BUSRQ1*/ BUSAK1* pair.

STEbus requests are level rather than edge triggered. A master, having asserted one of the bus request lines (either BUSRQ0* or BUSRQ1*) must continue to assert the line for as long as it wishes to gain control of the bus. Multiple masters may only be connected to a single bus request line (either BUSRQ0* or BUSRQ1*) if they will never simultaneously request the bus.

Having asserted a bus request line and detecting a corresponding bus acknowledge signal generated by the arbiter, a master may then assume control of the bus. Note that the arbiter will continue to assert the relevant bus acknowledge line for the entire period during which the current master continues to assert the corresponding bus request signal. The flow diagram of Figure 20 illustrates the STE control allocation sequence.

Data transfer protocol

All STEbus data transfers involve single bytes and an asynchronous interlocked handshake. This method of data transfer allows for a wide range of different system configurations and is instrumental in ensuring that the STEbus can provide effective communication between masters and slaves which operate at widely differing speeds.

Various types of data transfer sequence are allowed for including single

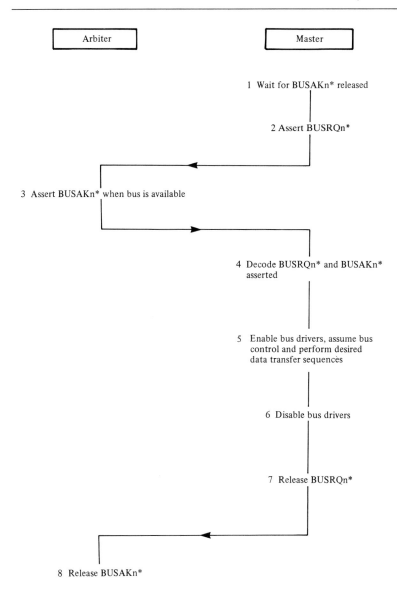

Figure 20 *Flow diagram showing the STE bus control allocation sequence*

data (read or write sequences), mixed data (read-modify-write), and block transfer (burst transfer sequences). These basic methods are shown in Figure 21. A vector-fetch sequence is employed when a master responds to a requesting board. This sequence is effectively a read sequence using a limited number of address lines.

The signals given in Table 8 are used during data transfer sequences.

56 STEbus hardware

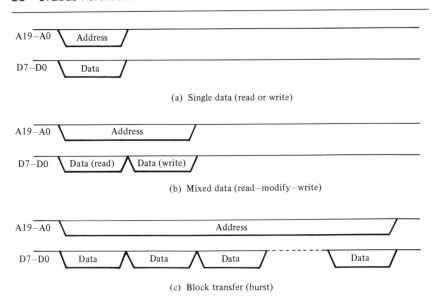

Figure 21 STEbus data transfer sequences
(a) Single data (read or write)
(b) Mixed data (read-modify-write)
(c) Block transfer (burst)

Table 8

Signal	Abbreviation	Source
Address lines	A0–A19	Master
Data lines	D0–D7	Master during write, slave during read and vector-fetch
Command modifiers	CM0–CM2	Master
Address strobe	ADRSTB*	Master
Data strobe	DATSTB*	Master
Data acknowledge	DATACK*	Slave
Transfer error	TRFERR*	Any board

General rules for data transfer

The IEEE-1000 specification includes the following seven rules governing data transfer using the STEbus:

1 A master shall not assert DATSTB* while ADRSTB* is released.
2 A slave shall not respond to DATSTB* while ADRSTB* is released.
3 A slave shall not assert TRFERR* and DATACK* simultaneously.
4 All address lines used in any transfer sequence shall be valid when

ADRSTB* is asserted, and shall remain valid until ADRSTB* is released.
5 All command lines shall be valid when DATSTB* is asserted, and shall remain valid until DATSTB* is released.
6 During read sequences all data lines shall be valid when DATACK* is asserted, and shall remain valid until DATSTB* is released.
7 During write sequences all data lines shall be valid when DATSTB* is asserted, and shall remain valid until DATSTB* is released.

Read sequence

During a read data transfer sequence, the master places the address of the targetted memory of I/O location on the address lines and, after a setup time during which the address lines become valid and after having ensured that DATACK* is released from the previous cycle, the master asserts ADRSTB*. The master also activates the command lines (CM0–CM2) to indicate the type of transfer according to Table 9.

Table 9

Type of operation	CM2	CM1	CM0
I/O read	1	0	1
Memory read	1	1	1

After a setup time during which the command signals become valid, the master asserts DATSTB* to indicate the presence of valid data on the command lines and that it is ready to accept data. The addressed slave then responds by enabling its bus drivers, placing requested data on the data lines.

After a further setup time during which the data lines become valid, the addressed slave asserts DATACK* indicating that data is available. In response to DATACK* the master accepts the data and releases DATSTB* indicating that the slave must remove its data from the data lines. Upon detecting either DATSTB* or ADRSTB* released, the slave disables its data bus drivers and releases DATACK* indicating that the sequence is complete.

Write sequence

During a write data transfer sequence the master places the address of the targetted memory or I/O location on the address lines and, after a setup time during which the address lines become valid and after having ensured that DATACK* is released from the previous cycle, the master

Table 10

Type of operation	CM2	CM1	CM0
I/O write	1	0	0
Memory write	1	1	0

asserts ADRSTB*. The master also activates the command lines (CM0–CM2) to indicate the type of transfer according to Table 10.

After a setup time during which the command signals become valid, the master asserts DATSTB* to indicate the presence of valid data on the command and data lines. The addressed slave then accepts the data and asserts DATACK* to indicate that the transfer may be terminated. In response to DATACK* the master releases ADRSTB* and DATSTB*. Upon detecting either DATSTB* or ADRSTB* released, the slave releases DATACK* indicating that the sequence is complete.

Read-modify-write sequence

A read-modify-write data transfer sequence is a sequence in which data transferred to the master is operated on in some way before being written back to the slave in one indivisible sequence.

The master first places the address of the targetted memory or I/O location on the address lines and, after a setup time during which the address lines become valid and after having ensured that DATACK* is released from the previous cycle, the master asserts ADRSTB*. The master also activates the command lines (CM0–CM2) to indicate the type of transfer according to Table 11.

Table 11

Type of operation	CM2	CM1	CM0
I/O read	1	0	1
Memory read	1	1	1

After a setup time during which the command signals become valid, the master asserts DATSTB* to indicate the presence of valid data on the command lines and that it is ready to accept data. The addressed slave then responds by enabling its bus drivers, placing requested data on the data lines.

After a further setup time during which the data lines become valid, the addressed slave asserts DATACK* indicating that data is available. In response to DATACK* the master accepts the data and releases DATSTB* indicating that the slave must remove its data from the data lines. Upon detecting DATSTB* released, the slave disables its data bus drivers and releases DATACK* indicating that the read transfer cycle is complete.

After operating on the data received and detecting DATACK* released, the master places a logic 0 on the CM0 line (while maintaining the logical state of the other command modifier lines) to signal a write (rather than a read) operation. After a setup time during which the data and CM0 lines become valid, the master asserts DATSTB* to indicate the presence of valid data on the command and data lines. The addressed slave then accepts the data and asserts DATACK* to indicate that the transfer may be terminated. In response to DATACK* the master releases ADRSTB* and DATSTB*. Upon detecting either DATSTB* or ADRSTB* released, the slave releases DATACK* indicating that the sequence is complete.

Vector fetch

In response to an attention request (ATNRQ*) signal, a master may perform a Vector Fetch operation. During a vector-fetch the master places the address (number) of the attention request (ATNRQ*) line on the three least significant address lines (A2–A0) and, after a setup time during which the A2 to A0 become valid, the master asserts ADRSTB*. The master also activates the command lines (CM0–CM2) to indicate the vector-fetch transfer according to Table 12.

Table 12

Type of operation	CM2	CM1	CM0
Vector-fetch	0	1	1

After a setup time during which the command signals become valid, the master asserts DATSTB* to indicate the presence of valid data on the command lines and that it is ready to accept data. The requesting device, upon decoding the address returned on A2–A0, then responds by enabling its bus drivers and placing its unique identifying vector on the data lines.

After a further setup time during which the data lines become valid, the requesting board asserts DATACK* indicating that vector data is available. In response to DATACK* the master accepts the vector data and releases ADRSTB* and DATSTB* indicating that the requesting board must remove its vector data from the data lines. Upon detecting either DATSTB* or ADRSTB* released, the requesting board disables its data bus drivers and releases DATACK* indicating that the sequence is complete.

Burst transfer sequences

STE also allows for a special form of read, write, or vector-fetch sequence in which the ADRSTB,* address and command modifier lines remain stable. During this period, multiple DATSTB* and DATACK* occur between the master and addressed slaves.

A slave board may be configured to participate in burst-mode transfer sequences by including logic which auto-increments or auto-decrements the on-board address on active edges of DATSTB*, or by the use of a FIFO (first-in, first-out) register array.

Transfer error

During a data transfer sequence a slave may assert TRFERR* when it wishes to indicate that a problem has arisen during the transfer. TRFERR* may be asserted (instead of DATACK*) at any time that ADRSTB* and DATSTB* are both active. The master then terminates the data transfer sequence in the usual way (i.e., as if DATACK* had been received) and the slave releases TRFERR* in the process. The master may then initiate whatever system-specific action may be appropriate.

System timeout

In order to prevent an STEbus system locking up in the event of a slave failing to respond (for example, when an addressed memory location is non-existent), the system controller should incorporate a timer to initiate abortion of an operation which fails due to lack of response from a slave.

The system controller should monitor all bus transfer sequences and assert TRFERR* at any time that the period from an active transition of DATSTB* to the subsequent active transition of DATACK* exceeds a designated maximum. In the event of a timeout occurring, the master should then terminate the aborted data transfer in the same manner as would be appropriate in the case of a TRFERR* signal generated by a slave. After the master releases DATSTB*, the system controller releases TRFERR* and the master may then take appropriate action to deal with the error.

Interboard signalling

The eight attention request lines are used for interboard signalling. These lines allow a board to alert one or more of the other boards present to the occurrence of some event which requires a change in flow or context of the current system activity. Such events include:

- Interrupts (i.e., conventional requests for service generated by slaves seeking the attention of a master).
- Requests for DMA transfer.
- Error conditions at system level (for example, power failure).
- Error conditions at board level (for example, parity errors).
- Any other system-specific or time-critical events.

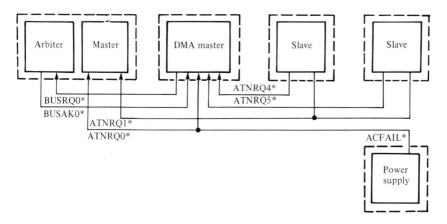

Figure 22 *Typical implementation of attention request lines in an STEbus system*

A typical implementation of the attention request lines in an STEbus system is shown in Figure 22. An attention request is initiated by a board asserting one of the eight attention request lines. Any board may be connected to any of the eight attention request lines, and multiple boards may be connected to the same attention request line. This feature allows requests or status flags to be broadcast on a global basis.

As mentioned earlier, attention request lines are prioritized with ATNRQ0* having the highest priority. Clearly the use to which attention request lines will be put will vary widely from system to system. Indeed, a measure of flexibility in the design of STE cards is required so that boards capable of responding to attention requests can have attention requests inhibited or reassigned. Furthermore, in the case of systems using attention request signals for interrupt handling as well as DMA, it is recommended that ATNRQ0*–ATNRQ3* should be used for interrupt requests while ATNRQ4*–ATNRQ7* should be employed for DMA requests.

Three types of responses are possible to an interrupt attention request: implicit, explicit, and local action. An *implicit* response involves a master first discovering which device is requesting attention (using one or more read or read-modify-write data transfer sequences) before acknowledging that the request has been granted. Depending upon the particular system configuration, the first stage in this process is not, of course, always necessary.

An *explicit* response involves the master initiating a vector-fetch data transfer sequence to determine which device is requiring attention. This process was discussed earlier under the general heading of data transfer protocol.

Unlike the other two types of response, a *local action* response does not involve any acknowledgement of the board initiating the response. It does, however, result in some form of activity within the board (or boards) receiving the attention request. A typical example of a local action response is the activity within a system when a power failure condition is broadcast via the attention request lines.

Electrical specifications

The STEbus backplane performs the following three functions:

1 Distribution of power supplies including:

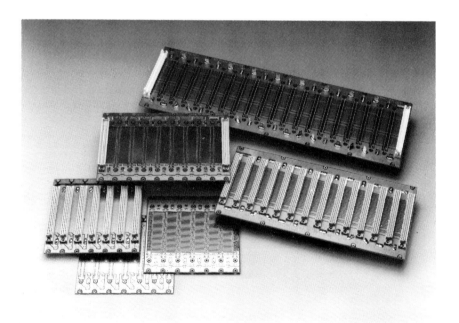

Figure 23 *A selection of STEbus backplanes with up to twenty-one card positions (courtesy of BICC-Vero Electronics Ltd)*

(a) Primary DC power rails using four pins on each connector designated +5 V and 0 V.
(b) Auxiliary supply rails of +12 V and −12 V.
(c) Provision for standby power.
2   Forty-nine signal lines which are normally terminated at both ends of the backplane.
3   Four AC 0 V (GND) lines which provide a common signal return for the signal lines in (2). These lines are embedded within the signal lines to act as backplane guard tracks.

The IEEE-1000 standard specifies a maximum length for any signal line (including connection to the terminating networks) of 500 mm (20 in). The backplane caters for a maximum of twenty-one boards at a preferred horizontal pitch of 20.32 mm (0.82 in).

Power supplies

The IEEE-1000 standard calls for the power supply characteristics shown in Table 13.

Table 13

Designation	Nominal value	Tolerance	Ripple	Maximum current per board	per system
$+V_{cc}$	+5 V	+5,%−2.5%	50 mV	4 A	50 A
+AUX V	+12 V	±5%	50 mV	1 A	4 A
−AUX V	−12 V	±5%	50 mV	1 A	4 A
+VSTBY	+5 V	+5%	50 mV	1 A	4 A

Power failure

Some means of orderly shutdown in the event of power failure is highly desirable in any process control system; the system can then take the necessary action to safeguard itself against corrupted data and all critical processes can be terminated.

As mentioned previously, the attention request line with highest priority (ATNRQ0*) should be used to warn the system of imminent power failure. The recommended protocol is then ATNRQ0* is asserted by a 'watchdog' device at least 4 ms *before* the primary power rail falls below its specified lower tolerance limit. The system controller asserts SYSRST* in response to the assertion of ATNRQ0.*

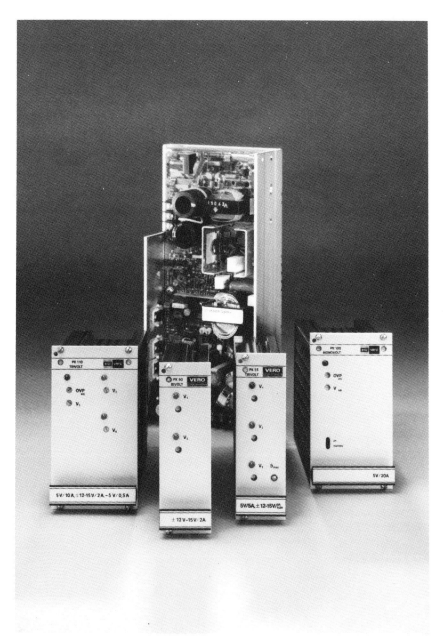

Figure 24 A selection of modular switched-mode power supplies for use with STE card frame assemblies (courtesy of BICC-Vero Electronics Ltd)

Bus drivers and receivers

STEbus drivers are required to have a minimum sink current capability of 24 mA at 0.5 V, though this voltage should ideally be increaed to 2.4 V. In order to improve noise margins, it is highly desirable for bus receivers to exhibit a hysteresis of at least 200 mV and thus Schmitt devices are preferred. This property is considered essential when dealing with the following bus synchronization signals: ADRSTB*; DATSTB*; DATACK*; TRFERR*; SYSCLK. It is also desirable for bus receivers to have some form of negative voltage clamping which will operate at input voltages of -1.4 V or less. The foregoing requirements are exemplified by such devices as the 74LS245 octal tristate bus transceiver.

Table 14 specifies the output circuitry of the bus drivers used in STE systems:

Table 14

Signal	Requirements
A0–A19	Tristate
D0–D7	Tristate
CM0–CM2	Tristate
ADRSTB*	Tristate
DATSTB*	Tristate
DATACK*	Open-collector or open-drain
BUSRQ0*–BUSRQ1*	Open-collector or open-drain
TRFERR*	Open-collector or open-drain
ATNRQ0*–ATNRQ7*	Open-collector or open-drain
SYSRST*	Open-collector or open-drain
SYSCLK	Totem-pole
BUSAK0*–BUSAK1*	Totem-pole

Signal voltages and bus logic levels

As with most TTL compatible bus systems, a voltage of less than 0.8 V should be recognized by a receiver as logic 0 while a voltage of greater than 2.0 V is taken to represent a logic 1, as shown in Figure 25.

Termination networks

When more than five slot positions are fitted to the backplane, the backplane signal lines should normally be terminated at each end. When there are five or less slot positions it is usually permissible to terminate the backplane at one end only.

Bus terminations should represent a constant voltage source of 2.8 V ± 10% having an internal resistance of 270 Ω ±5% (as shown in Figure 26).

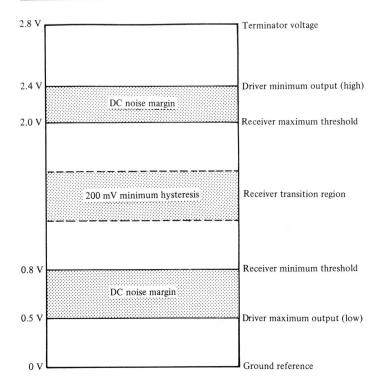

Figure 25 *Voltage and logic levels in an STE bus system*

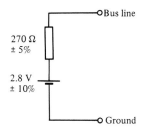

Figure 26 *Equivalent circuit of an STE bus termination*

This may be realized using a network of resistors connected to the primary +5 V rail (as shown in Figure 27) or by means of an active arrangement using a constant voltage source (as shown in Figure 28). In either case, decoupling capacitors of suitable value (typically 100 nF) should be fitted in close proximity to the resistor networks.

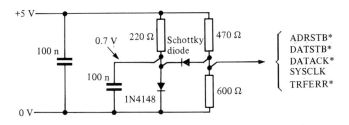

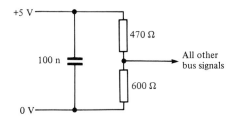

Figure 27 *Practical realization of Figure 26 using passive resistor networks*

The STEbus connector

The connector specified for use with STE boards is a 64-pin male connector utilizing rows a and c (the inside row is not used). This connector is specified in IEC 603–2 and the corresponding female connector is specified for the STE backplane.

The STEbus connector pin-out is shown in Figure 30. The function of the signals present are as follows:

D0 to D7	Eight data lines.
A0 to A19	Twenty address lines.
ADRSTB*	Address strobe. This line is taken low to indicate that a valid address has been placed on the bus.
DATSTB*	Data strobe. This line is taken low to indicate that valid data has been placed on the bus.
CM0 to CM2	Command modifiers which indicate the current type of bus cycle.
BUSRQ0*–1*	Bus request lines. These lines are taken low when a potential bus master wishes to gain access to the bus.
BUSAK0*–1*	Bus acknowledge lines. These lines are taken low to indicate that the bus request has been granted. A potential bus master may only drive the bus when it has received an acknowledge signal on the bus request line.

68 STEbus hardware

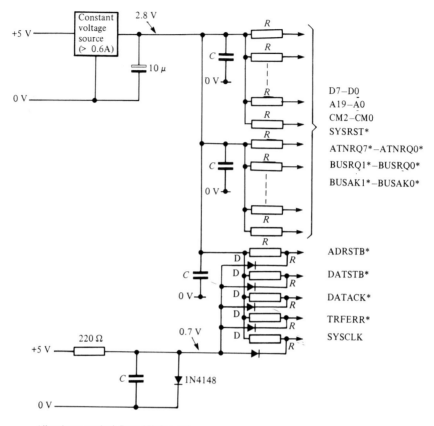

All resistors marked R are 270 Ω ± 5%.
All capacitors marked C are 100 nF.
All diodes marked D are Schottky types.

Figure 28 *Practical realization of Figure 26 using an active constant voltage source*

DATACK* This handshake line is asserted by a bus slave on a write cycle in order to indicate that it has accepted data or, on a read cycle, to indicate that its data is valid.
TRFERR* A bus slave asserts this signal instead of DATACK* if an error is detected.
ATNRQ0*–1* Attention request/interrupt lines. (ATNRQ0* has the highest priority.)
SYSCLK 16 MHz system clock.
SYSRST* System reset.

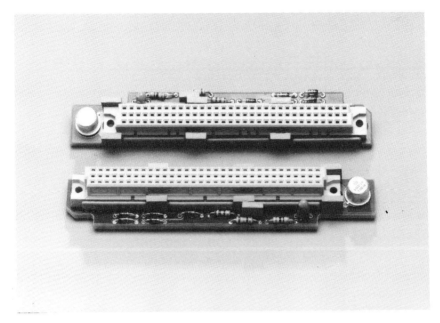

Figure 29 *Active bus termination networks (courtesy of BICC-Vero Electronics Ltd)*

STEbus boards

Having spent some time discussing the STEbus standard it is worth taking a look at some typical STEbus cards. A huge range of STE boards is currently available including a wide variety of processor cards, I/O boards, memory cards, system watchdogs, floppy disk controllers and IEEE-488 interfaces.

Typical STEbus processors

STEbus processors invariably comprise a single board computer containing microprocessor CPU, ROM, RAM and bus interfacing hardware. Since all of this must be contained on a standard eurocard (measuring 100 mm × 160 mm) the packing density is necessarily high. Indeed, one of the most popular STEbus processors uses over thirty chips and four of these are 40-pin DIL types.

A typical STEbus processor is shown in outline block schematic form in Figure 32. The processor is a Z80-CPU operating at clock frequency of 4 MHz. The basic system clock generator operates at 16 MHz and is controlled by a 16 MHz quartz crystal. The 16 MHz clock is successively divided to provide the following clock signals:

70 STEbus hardware

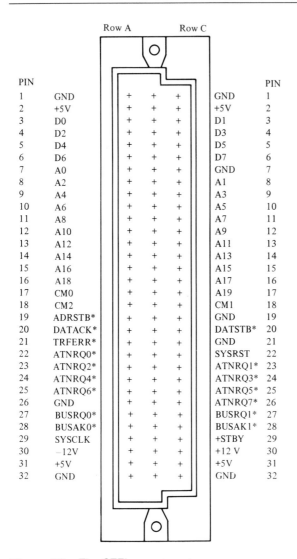

PIN			PIN
1	GND	GND	1
2	+5V	+5V	2
3	D0	D1	3
4	D2	D3	4
5	D4	D5	5
6	D6	D7	6
7	A0	GND	7
8	A2	A1	8
9	A4	A3	9
10	A6	A5	10
11	A8	A7	11
12	A10	A9	12
13	A12	A11	13
14	A14	A13	14
15	A16	A15	15
16	A18	A17	16
17	CM0	A19	17
18	CM2	CM1	18
19	ADRSTB*	GND	19
20	DATACK*	DATSTB*	20
21	TRFERR*	GND	21
22	ATNRQ0*	SYSRST	22
23	ATNRQ2*	ATNRQ1*	23
24	ATNRQ4*	ATNRQ3*	24
25	ATNRQ6*	ATNRQ5*	25
26	GND	ATNRQ7*	26
27	BUSRQ0*	BUSRQ1*	27
28	BUSAK0*	BUSAK1*	28
29	SYSCLK	+STBY	29
30	−12V	+12 V	30
31	+5V	+5V	31
32	GND	GND	32

Figure 30 *The STEbus connector*

1 8 MHz for the dynamic RAM controller.
2 4 MHz for the CPU and RS-232C serial interface.
3 2 MHz for the disk controller.

The system 16 MHz clock is also fed to the STEbus for use by slave cards. Since only one 16 MHz bus clock signal should be present and more than one processor card may be fitted, a facility for disabling this output is incorporated in the form of a link on the board.

Figure 31 *A typical STEbus processor having a Z80 CPU, 64K of RAM, and a floppy disk controller (courtesy of Arcom Control Systems Ltd)*

The dynamic RAM controller provides the multiplexed data and active low row address and column address select ($\overline{RAS}$ and $\overline{CAS}$) signals for the eight 64 K × 1 bit dynamic RAM chips. The disk controller is a dedicated LSI device and the serial interface is a programmable serial interface controller. The serial interface incorporates the necessary level shifting to implement a full-specification RS-232C interface.

The data and address buses are buffered from the STE backplane by means of two octal drivers (in the case of the address bus) and one octal transceiver (in the case of the data bus). All three of these devices have tristate outputs and can thus be isolated from the external bus when required.

A typical STE digital I/O board

A typical STE digital I/O board is shown in the outline block schematic of Figure 33. The digital I/O board has 40 I/O lines which are accessible in groups of eight, and four of these groups have TTL buffers. The fifth group is unbuffered and can either be used as I/O lines or as status lines for the other groups. Each of the buffered groups can be set by software to be either input or output, and the buffer output can be enabled or disabled by

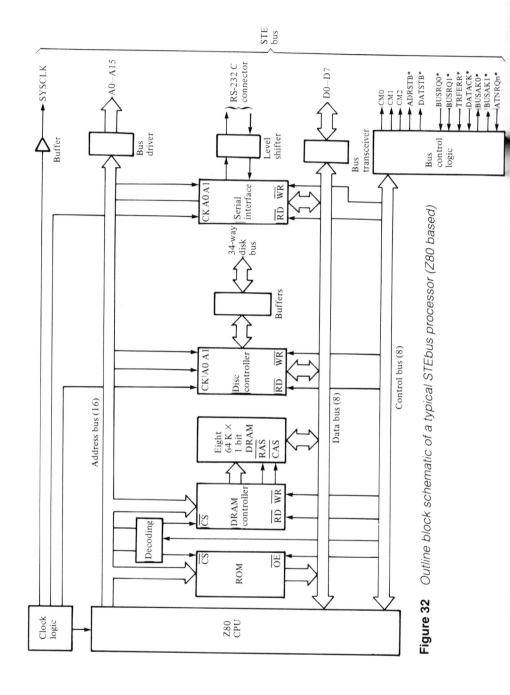

Figure 32 Outline block schematic of a typical STEbus processor (Z80 based)

STEbus boards 73

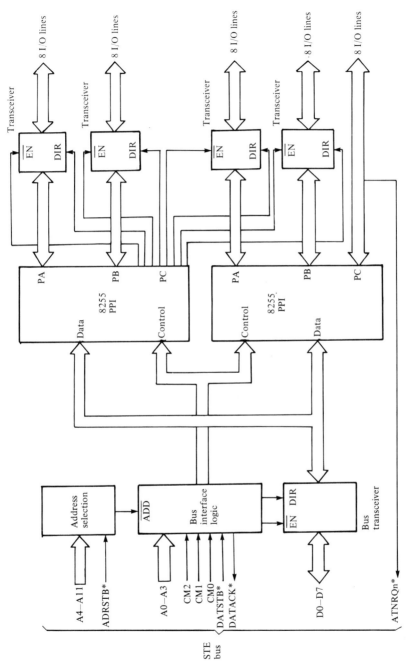

Figure 33 Outline block schematic of a typical STE digital I/O board

software. The unbuffered group can be used in a number of ways, with various bits designated for input, output, handshake or interrupt.

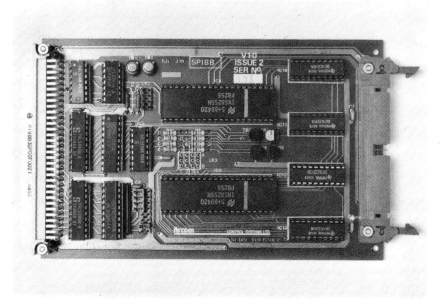

Figure 34 *A typical STEbus digital I/O board based on two 8255 PPI devices and providing forty parallel I/O lines (courtesy of Arcom Control Systems Ltd)*

One of the buffered groups, when used with an input line from the unbuffered group, can act as a latched input which generates an interrupt. This interrupt can be jumpered to one of four attention request lines on the STEbus.

It is often necessary to set output of a slave card to a defined state on power up or reset without creating transitory states (glitches) as a system is initialized. This is achieved by delaying the enabling of the TTL transceivers used as output buffers until after the parallel I/O devices have been configured.

The digital I/O board is mapped to the STE I/O space and it appears as eight I/O locations which can be set to appear within any one of 256 address groups. A jumper option is provided on the board so that the four most significant address bits of the 12-bit I/O address are ignored. This makes the task of I/O programming much simpler when using an 8-bit processor which has an instruction set containing instructions of the form: OUT (port), accumulator and IN accumulator, (port).

The digital I/O board uses two 8255 PPI (programmable peripheral interface) devices. Digital inputs and outputs are derived from a 50-way ribbon-cable connector which conforms to the commonly used STE

STEbus boards 75

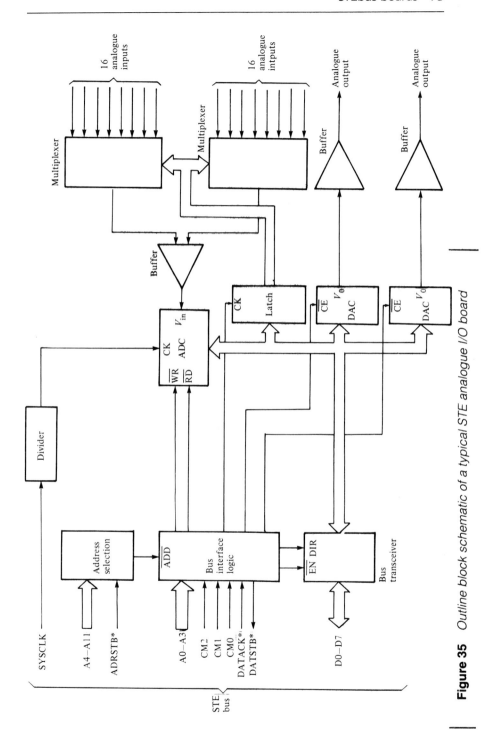

Figure 35 Outline block schematic of a typical STE analogue I/O board

digital I/O pin configuration (see Appendix 4 for details). This connector is fitted to the edge of the board opposite to that occupied by the 64-way DIN-41612 bus connector.

A typical STE analogue I/O board

The outline schematic of a typical analogue I/O board is shown in Figure 35. The board provides eight inputs and two outputs. The board multiplexes eight differential (or sixteen single ended) input channels into a high-speed operational amplifier which drives an 8 bit successive approximation analogue to digital converter (ADC). Two separate digital to analogue converters (DAC) are provided on the board, each having its own high-speed operational amplifier output buffer.

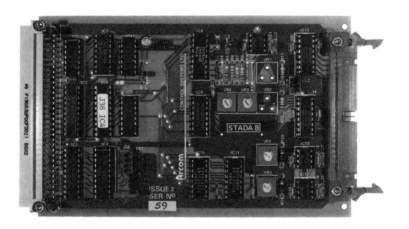

Figure 36 *A typical STEbus analogue I/O board providing eight analogue inputs and two analogue outputs and offering 8-bit resolution (courtesy of Arcom Control Systems Ltd)*

The board is mapped to three I/O locations which can be set to one of 256 address groups within the 4 K I/O space. As with the digital I/O board, a jumper option may be set to ignore the four most significant bits of the I/O address.

Analogue inputs and outputs are derived from a 50-way ribbon-cable connector which conforms to the commonly used STE analogue I/O pin configuration. This connector is fitted to the edge of the board opposite to that occupied by the 64-way DIN 41612 bus connector (see Appendix 5 for details).

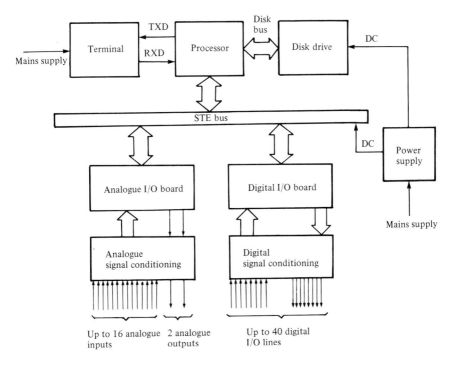

Figure 37 *Complete STEbus configuration using the boards depicted in Figures 32, 33 and 34*

A typical STEbus configuration

A typical STEbus configuration (using the three boards previously described) is shown in Figure 37. A single Z80 processor card is used in conjunction with two slave I/O boards; a digital I/O board and an analogue I/O board. The system provides forty digital I/O lines, eight analogue inputs and two analogue outputs.

Program and data storage is provided by means of a disk drive and the system accepts commands via a serial RS-232C link from a terminal or host microcomputer operating in *terminal emulation mode*. The backplane houses a Eurocard mounted power supply rated at 85 W with the following outputs:

- +5 V at 8 A maximum (primary DC power to bus and +5 V to disk drive).

- +12 V at 3 A maximum (disk drive power).
- +12 V at 1.5 A maximum (auxiliary power to bus).
- −12 V at 1 A maximum (auxiliary power to bus).
- −5 V at 0.5 A maximum (spare).

I/O signal conditioning is required to interface input and output transducers to the system. This takes the form of several signal conditioning boards which are linked to the two I/O cards by separate 50-way ribbon cables. Each of the signal conditioning cards is fitted with a 48-way DIN 41612 connector which mates with a cable terminator on which the transducer wiring is terminated.

4
Operating systems

Anyone who has made passing use of a microcomputer system will be aware of at least some of the facilities offered by its operating system. Such an awareness is developed by means of the interface between the operating system and the user; the system generates prompts and messages and the user makes an appropriate response.

Many of the functions of an operating system (like those associated with disk filing) are obvious. Others, however, are so closely related to the machine's hardware that the average user remains blissfully unaware of them. This, of course, is as it should be. As far as most end-users of computer systems are concerned, the operating system provides an environment from which it is possible to launch and run applications programs and to carry out elementary maintenance of disk files. Here, the operating system would perhaps be better described as a microcomputer resource manager.

In the context of software development, on the other hand, the operating system takes on a new significance. The software developer needs to interact with the operating system at a much lower level than the applications user, hence the operating system becomes crucial in making effective use of the resources offered by the system. Furthermore, a range of system specific utility programs will be required (including such items as editors, assemblers, linkers, loaders and debuggers). These software tools must work together with the operating system to provide an environment which will facilitate effective software development.

This chapter outlines the facilities provided by an operating system which are of particular interest to the software developer. The details and examples given relate to the ever-popular CP/M-Plus (CP/M 3) operating system. However equivalent functions and commands will be found in virtually every operating system.

Unfortunately, it is not possible to provide an exhaustive description of the facilities offered by any operating system in a single chapter. It thus remains for readers to obtain full documentation of the operating system they intend to use and spend some time in familiarizing themselves with the range of facilities offered and the syntax used for commands. Time spent in getting to know an operating system (including its peculiarities and foibles) can be instrumental in saving time and avoiding a variety of pitfalls at a later stage.

CP/M

Most microcomputer operating systems can trace their origins in the original Control Program for Microprocessors (CP/M) developed by Gary Kildall as a software development environment for the Intel 8080 microprocessor. In its original form CP/M was supplied on an 8-inch IBM format floppy disk. CP/M was subsequently extensively developed and marketed by Digital Research and now exists in a variety of forms including those for use with Intel and Motorola 16-bit microprocessor families.

CP/M rapidly became the *de facto* operating system for most 8080, 8085 and Z80 based 8-bit microcomputer systems. Furthermore, the wealth of business software to run under CP/M has guaranteed its long-term survival in the face of strong competition from arguably better and more user-friendly operating systems.

The following major variants of CP/M are in existence:

CP/M-80	The original 8080-based operating system.
CP/M-2.2	Improved and enhanced version of CP/M.
CP/M-Plus	Also known as CP/M-3, enhanced CP/M with Z80 macro-assembler, faster disk thoughput and banked memory facilities.
CP/M-86	CP/M for 8086 family microprocessors.
CP/M-68K	CP/M for 68000 family microprocessors.
Concurrent CP/M	CP/M which permits time-sharing of the processor between two or more tasks.
MP/M-86	A very much enhanced multi-processing operating system for the 8086 family microprocessors.

CP/M-Plus

CP/M-Plus has proved to be one of the most popular operating systems used for software development in conjunction with simple low-cost industrial process control systems. It provides a huge range of commands and is supplied in versions which support banked and non-banked memory. In the banked version, CP/M-Plus offers extended line editing, password protection of files and extended error messages.

The non-banked version of CP/M-Plus requires 8.5 K of RAM, plus additional space for a customized BIOS (basic input/output system). This element of CP/M contains hardware specific I/O routines and therefore must be adapted for each different environment on which the system is to run.

CP/M-Plus is loaded from disk during initialization of the system (i.e., on power-up or after a hardware reset) by means of a short program known as the *cold boot loader*. This program passes control to a system loader (CPMLDR) which completes the loading process.

The various components of CP/M-Plus are shown in the memory map of Figure 38. The hardware-dependent BIOS module (which defines the low-level interface with a particular system) is located in the highest block of memory. Immediately below the BIOS module is the basic disk operating system (BDOS). This module provides a standard operating system environment which is independent of a particular hardware configuration and it is discussed at greater length later in this chapter.

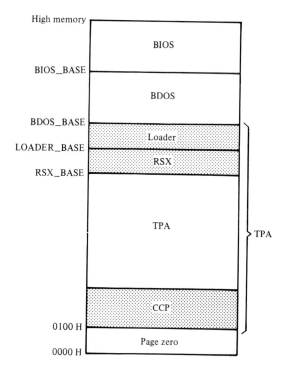

Figure 38 *Basic CP/M-Plus memory map*

The CP/M-Plus LOADER module is responsible for loading programs when required by the console command processor (CCP) and by utility programs. Like the CCP, the LOADER is not usually resident when transient programs execute. When the LOADER is resident, however, transient programs may access the module using BDOS Function call 59.

As well as the LOADER module, one or more resident system extensions (RSX) may be present. These modules are used to extend or modify normal functions of the operating system. The transient program area (TPA) is the name given to the region of memory in which transient programs (i.e., the user's own applications) execute. The normal default address for the start of the TPA is 0100H. The console command processor (CCP) is a system program which is located in low-memory (above PAGE ZERO) when resident. The CCP is used to provide the all-important interface between the operating system and the user.

PAGE ZERO is the name given to the lowest 256 bytes of memory (addresses 0000H to 00FFH). This area is reserved for system use and provides the interface to the BDOS module from the CCP and transient programs as well as a storage area for system parameters.

Loading the operating system

CP/M-Plus is supplied on a distribution diskette (invariably either 3.5 in or 5.25 in) and the operating system must be loaded into memory before it can be used. The system should be switched on and the system disk inserted into the *default* drive (drive A on a multidrive system). Having closed the drive door, the system reset (restart) button should be depressed. The operating system will then automatically be loaded (booted).

After successful loading, the title and version of the operating system will be displayed on the console. This message is then followed by a prompt which gives the currently selected drive (A) and indicates that the system is ready to accept a command.

Note that the currently selected drive is the default drive in that, unless another drive is specified within a command, all drive related commands are taken to refer to this drive.

Making back-up copies of system disks

Before making use of the software supplied on a distribution disk, it is wise to make at least one back-up copy. Having done so, the distribution disk should be safely stored away. The CP/M-Plus utilities COPYSYS or SYSCOPY should be used to copy the operating system. The procedure for making a back-up varies depending upon whether the system is a single or multidrive.

Multidrive systems

When two (or more) drives are available, the system should be booted (using the method previously described) with the distribution disk placed in drive A (the default drive). After booting, a new (blank) disk should then be placed in drive B.

The disk in drive B should then be formatted using the DISCFORM (or equivalent) utility. The following command should be entered:

A>DISCFORM B:

Formatting will take several minutes (with a message showing the current track being formatted. The formatting process may be prematurely aborted using CTRL-C.

Now copy the distribution diskette to the newly formatted disk using the command:

A>DISCCOPY A: B:

After a few more minutes, the copying process will be completed and the distribution disk (in drive A) should be removed and stored away safely. The back-up system disk can then be removed from drive B and inserted into drive A.

Single drive systems

When only a single drive is available, the system should be booted (using the method previously described) with the distribution disk placed in drive A (the default drive). After booting, the following command should be entered:

A>DISCFORM A:

(Note that the system will not accept the DISCFORM command without a drive specification.)

A message will then appear on the screen reminding users to remove the system disk from the drive. The distribution disk should then be removed and replaced by a blank disk before pressing the RETURN key to start the formatting process.

Formatting will take several minutes (with a message showing the current track being formatted). The formatting process may again be prematurely aborted using CTRL-C.

When the formatting process has been completed, the user is prompted to remove the newly formatted disk and replace it with the system disk. Having placed the distribution disk back in the drive, enter the following command:

A>DISCCOPY A: A:

The system will then repeatedly prompt the user to swap the two disks in the single drive. The *source* disk referred to in the system prompts is the distribution (master) disk while the *destination* disk is the back-up (copy). After several minutes of disk swapping, the process will be complete (with the destination disk still in the drive) and the user can store away the distribution disk. The back-up copy then becomes the user's working copy of the operating system and is left in the drive.

Logical devices

CP/M-Plus supports five logical devices: CONIN, CONOUT, AUXIN, AUXOUT and LST, and a number of system dependent physical devices which are recognized by names such as CON, AUX and CRT. To distinguish between logical and physical devices, the former are normally terminated by a colon.

It should be obvious that CONIN and CONOUT are jointly associated with the keyboard and VDU of the console (CON) device. Similarly, the KEYBOARD device is normally synonymous with CONIN. AUXIN and AUXOUT usually refer to the auxiliary serial device (AUX) whereas the LST device is conventionally linked to the printer, LPT.

The distinction between logical devices and physical peripherals may seem unnecessarily cumbersome but it is intrumental in allowing CP/M to redirect data. This feature can be extremely useful when, for example, output normally destined for the printer is to be redirected to an auxiliary serial port. The DEVICE command can be used both to display and modify the assignment of logical devices to physical peripherals.

CP/M-Plus commands

CP/M-Plus responds to command lines typed at the console and terminated with a RETURN or ENTER character. A command line is thus composed of a command keyword, an optional command tail, and a RETURN keystroke. The command keyword identifies the command (or program) to be executed. The command tail can contain extra information relevant to the command, such as a filename or other parameters. The command line is terminated using the RETURN or ENTER key.

As an example, the following command can be used to display a directory of all command files (i.e., those with a COM extension) indicating the size of each:

DIR [SIZE] *.COM   (see Figure 39)

```
A>DIR [SIZE] *.COM

Scanning Directory...

Sorting Directory...

Directory For Drive A:    User  0

A: CCP      COM   4k : COPY     COM   4k : CPMLDR  COM    4k
A: DATE     COM   4k : DEVICE   COM   8k : DIR     COM   16k
A: DISCCOPY COM   4k : DISCFORM COM   4k : DUMP    COM    4k
A: ED       COM  12k : EPROM    COM   4k : ERASE   COM    4k
A: GENCOM   COM  16k : GENCPM   COM  24k : GET     COM    8k
A: HELP     COM   8k : HEXCOM   COM   4k : INITDIR COM   32k
A: LIB      COM   8k : LINK     COM  16k : MAC     COM   12k
A: PATCH    COM   4k : PIP      COM  12k : PUT     COM    8k
A: RENAME   COM   4k : RMAC     COM  16k : SAVE    COM    4k
A: SET      COM  12k : SETDEF   COM   4k : SHOW    COM   12k
A: SID      COM   8k : SUBMIT   COM   8k : SYSCOPY COM    4k
A: TYPE     COM   4k : XREF     COM  16k : ZSID    COM   12k

Total Bytes     =    328k  Total Records =    2165  Files Found =    36
Total 1k Blocks =    287   Used/Max Dir Entries For Drive A:    90/ 128
```

Figure 39 *Typical response to the CPlM/Plus DIR [SIZE]*.COM command*

Note that, in this example and the examples which follow, we have omitted the prompt generated by the system (indicating the current drive), and the ENTER or RETURN character which terminates the command.

It should be noted that the command line can be entered in any combination of upper-case or lower-case characters. CP/M-Plus converts all letters in the command line to upper-case before interpreting them. Furthermore, while a command line generally immediately follows the system prompt, CP/M-Plus permits spaces between the prompt (>) and the command keyword.

As characters are typed at the keyboard, the cursor moves to the right in order to indicate the position of the next character to be typed. Depending upon the keyboard used, a BACKSPACE, RUBOUT, or DELETE key can be used to delete the last entered character and move the cursor backwards one character position. If a keyboard is not fitted with such a key then the combination of CONTROL and H (CTRL-H) may be used instead. Note that the CONTROL (CTRL) key should be depressed first, followed by the H key. Various CP/M-Plus control characters are shown in Table 15.

Intrinsic and extrinsic commands

A distinction is made between CP/M commands which relate to the resident part of the operating system (intrinsic commands) and those which involve other utility programs (extrinsic commands). Intrinsic commands are executed immediately whereas extrinsic commands require the loading of transient utility programs from disk and hence there is a short delay before the command is acted upon.

Table 15 CP/M-Plus control characters

Character	Function
CTRL-C	Terminates the current program and redisplays the system prompt provided the cursor is at the beginning of the command line.
CTRL-E	Forces a physical carriage return but does not send the command line to CP/M-Plus. Moves the cursor to the beginning of the next line without erasing previous input.
CTRL-H	Deletes a character and moves the cursor left one character position.
CTRL-I	Moves the cursor to the next tab stop (i.e., has the same effect as TAB). Tab stops are automatically set at each eighth column.
CTRL-J	Sends the command line to CP/M-Plus and returns the cursor to the left of the current line (i.e. has the same effect as RETURN).
CTRL-M	Same as CTRL-J.
CTRL-P	Echoes all console activity to the printer. A subsequent CTRL-P is required to disable this facility.
CTRL-R	Places a # at the current cursor location, moves the cursor to the next line, and displays any partial command typed thus far.
CTRL-S	Disables screen scrolling. Scrolling can be subsequently re-enabled by means of a further CTRL-S.
CTRL-U	Discards all characters in the command line, places a # at the current cursor position, and moves the cursor to the next command line.
CTRL-X	Discards all the characters in the command line and moves the cursor to the beginning of the current line.

In the case of extrinsic commands, CP/M-Plus checks only the command keyword. Any command tails are passed to the utility program without checking. A command tail cannot contain more than 128 characters.

At this point we should perhaps mention that CP/M-Plus only recognizes command keywords which are correctly spelled! Even an obvious typing error will result in the non-acceptance of the command and the system will simply echo the faulty command line followed by a question mark.

As an example, suppose the user is attempting to format a disk but types DISKFORM instead of DISCFORM as follows:

A>DISKFORM

```
A>DIR B : [FULL]

Scanning Directory...

Sorting Directory...

Directory For Drive B:  User  0

Name       Bytes   Recs   Attributes   Prot      Update          Create
------     -----   ----   ----------   ----   --------------   --------------
BDOS3    SPR  12k    77   Dir RO       None   06/19/86 13:14   06/19/86 13:14
BIOS3    SPR   4k    23   Dir RO       None   11/07/85 11:03   11/07/85 11:03
BIOSKRNL ASM  16k   125   Dir RW       None   11/07/85 11:05   11/07/85 11:05
BNKBDOS3 SPR  16k   106   Dir RO       None   06/19/86 13:15   06/19/86 13:15
BNKBIOS3 SPR   4k    25   Dir RW       None   11/07/85 11:03   11/07/85 11:03
BOOT     ASM   4k    32   Dir RO       None   11/07/85 11:07   11/07/85 11:07
BOOTB    REL   4k     6   Dir RW       None   11/07/85 10:49   11/07/85 10:49
BOOTN    REL   4k     5   Dir RO       None   11/07/85 10:49   11/07/85 10:49
CCP      COM   4k    25   Sys RO       None   11/07/85 10:44   11/07/85 10:44
CHARIO   ASM   4k    23   Dir RW       None   11/07/85 11:06   11/07/85 11:06
CHARIO   REL   4k     3   Dir RO       None   11/07/85 10:50   11/07/85 10:50
COPY     COM   4k    14   Sys RO       None   11/07/85 10:44   11/07/85 10:44
CPM3     LIB   4k    32   Sys RO       None   11/07/85 10:50   11/07/85 10:50
CPM3     SYS  12k    84   Dir RW       None   12/15/82 00:00   12/15/82 00:00
CPMLDR   COM   4k    29   Sys RO       None   11/07/85 11:08   11/07/85 11:08
CPMLDR   REL   4k    23   Dir RO       None   11/07/85 10:50   11/07/85 10:50
DATE     COM   4k    22   Sys RO       None   11/07/85 10:45   11/07/85 10:45
Press RETURN to Continue
```

Figure 40 *Typical response to the CP/M-Plus DIR: [FULL] command*

88 Operating systems

The system responds with the message:

DISKFORM?

indicating that the command is unknown.

We shall now briefly examine the function of each of the commands provided in CP/M-Plus. Examples have been included where these may help to clarify the action of a particular command. Readers itending to use CP/M-Plus should refer to the *CP/M 3 User's Guide* supplied by Digital Research. This document provides the full syntax and range of options associated with each command.

Intrinsic CP/M commands

Table 16

Command	Function
DIR	The DIR command displays the names of all non-system files in the directory. Variations of the command allow the user to specify the drive to be searched and the types of files to be displayed. Further options govern the format of the directory display. *Examples* DIR Displays all files on the default drive. DIR B: Displays all files on the disk in drive B. DIR *.BAS Displays all files with a BAS extension on the current drive. DIR C:TRAN.* Displays all files named TRAN (regardless of their type or extension) present on the disk in drive C. DIR B: [FULL] Displays a full directory (including file attributes and date/time information) for the disk in drive B (see Figure 40).
DIRSYS	The DIRSYS (or DIRS) command displays the names of all system files in the directory. *Example* DIRS C:*.COM Displays all system files of type COM on the disk in drive C.
ERASE	The ERASE command is used to erase a filename from the directory and release the storage space occupied by a file. ERASE may be used with the * wildcard character in order to erase all files for which a match occurs. ERASE may be abbreviated to ERA. *Examples* ERASE PROG1.ASM Erases the file named PROG1.ASM from the disk placed in the current default drive. ERASE B:TEMP.DAT Erases the file named TEMP.DAT from the disk placed in drive B. ERA C:*.COM Erases all files having a COM extension present on the disk in drive C.

CP/M-Plus commands 89

	ERA A:PROG1.* Erases all files named PROG1 (regardless of their type extension) present on the disk currently in drive A.
RENAME	The RENAME command allows the user to rename a disk file. RENAME may be used with the * wildcard character in order to rename all files for which a match occurs. RENAME may be abbreviated to REN. If file specifications are omitted from the RENAME command, the system generates appropriate prompts. *Examples* RENAME PROG2.ASM=PROG1.ASM Renames PROG1.ASM to PROG2.ASM on the disk placed in the current default drive. REN A:HELP.DAT=TEXT.TXT Renames the file TEXT.TXT to HELP.DAT on the disk placed in drive A. REN B:CONTROL.*=PROG1.* Renames all files with name PROG1 (regardless of type extension) to CONTROL (with identical extensions) found on the disk placed in drive B.

TYPE	The TYPE command allows the user to display the contents of an ASCII (text) file on the console screen. The TYPE command can be used with options which enable or disable paged mode displays. The CTRL-S key combination may be used to halt the display and CTRL-Q used to restart. CTRL-C may be used to abort the execution of the TYPE command and exit to the system.
	   The file may be simultaneously echoed to the printer by means of CTRL-P which should be used before issuing the TYPE command. A second CTRL-P can be issued to disable the echo facility.
	   The TYPE command may be abbreviated to TYP.
	Example
	TYPE B:PROG1.ASM Will display the contents of a file called PROG1.ASM stored on the disk placed in drive B. The file will be sent to the console screen.

USER	The USER command can be used to set the current user number (in the range 0 to 15). USER may be abbreviated to USE.

CP/M utility commands

Important note: Unlike intrinsic commands, these commands will not function unless the appropriate utility program is resident on the system disk.

Table 17

Command	Function
COPY	The COPY command can be used to transfer a file from one disk to another using the same or a different filename. The COPY command is effective when the user has only a single drive. The COPY command must be followed by one or two file

```
A>DEVICE

Physical Devices:
I=Input,O=Output,S=Serial,X=Xon-Xoff
CRT     4800    IOS     LPT     1200    IOS     KBD     NONE    I
CENT    NONE    O               GRA     NONE    IO

Current Assignments:
CONIN:   = CRT
CONOUT:  = CRT
AUXIN:   = LPT
AUXOUT:  = LPT
LST:     = LPT

Enter new assignment or hit RETURN
```

Figure 41 *Typical response to the CP/M-Plus DEVICE command*

specifications. When only a single file specification is given, the command makes a single drive copy of a file. The copied file takes the same filename as the original and the user is prompted to insert the source and destination disks at the appropriate point. Where both source and destination file specifications are included, the file is copied to the specified drive and the copy takes the specified name. Where only a destination drive is specified (i.e., the destination filename is omitted) the COPY command copies the file to the specified drive without altering the filename.

Examples
COPY A:ED.COM   Copies the file ED.COM present on the disk in drive A to another disk inserted in the same drive. The COPY utility generates prompts during the process.
COPY A:ED.COM B: Copies the file ED.COM present on the disk in drive A to the disk present in drive B. The copy will be given the name ED.COM.

COPYSYS	The COPYSYS command creates a new boot disk by copying the CP/M system files. Disks used must have the same format. (Note that the COPYSYS utility may be replaced by the SYSCOPY utility, see page 82).
DATE	Sets or displays the date and time. During initialization, the date and time are set by default to the creation date of the CP/M system. The format for a time specification is MM/DD/YY HH:MM:SS. Where MM is the month (in the range 1 to 12); DD is the day (in the range 1 to 31); YY is the two-digit year (relative to 1900); HH is the hour (in the range 0 to 23); MM is the minute (in the range 0 to 59); and SS is the second (in the range 0 to 59) The validity of a date/time entry is checked by the system and the day of the week is computed from the validated date. *Examples* DATE Displays the current date and time (including the day of the week). DATE 01/02/89 10:30:00 Sets the date/time to 10.30 a.m. on 2 January 1989.

CP/M-Plus commands 91

DEVICE The DEVICE command has several functions including:
- Displaying the current assignment of logical devices together with the names of associated peripherals.
- Assigning logical devices to peripherals.
- Changing device driver protocol and baud rates.
- Displaying or changing the current console screen parameters.

Examples
DEVICE Displays the physical devices and current assignments of the logical devices in the system (see Figure 41).
DEVICE CONOUT:=LPT Directs console output to the printer by assigning CONOUT to LPT.
DEVICE LPT[XON,9600] Sets the XON/XOFF protocol for the physical device LPT and sets the baud rate to 9600.
DEVICE CONSOLE[COLUMNS=40,LINES=16] Sets the console page width to 40 column and the page length to 16 lines.

DISCFORM The DISCFORM command formats the disk in the specified drive.
Example
DISCFORM B: Format the disk in drive B.

DUMP The DUMP command displays a file in ASCII and hexadecimal format.
Example
DUMP B:TEMP.DAT Displays the file called TEMP.DAT present on the disk in drive B.

ED The ED utility allows the user to create and modify text files. ED operates as a line editor (see Chapter 5) and the command is followed by an input file specification and an (optional) output file specification.
Examples
ED PROG2.BAS Creates the file PROG2.BAS (if not already present) on the current drive. The editor prompt :* then appears to indicate that ED is ready to accept a command.
ED ONE.ASM B: Creates (or opens) the file ONE.ASM for editing using the current default drive. Edited output is sent to a disk file with the same name in drive B.

GENCOM The GENCOM command invokes a utility which creates a program (COM) file with attached resident system extension (RSX) files. The program file is given a special header to indicate that RSX loading is required.
Example
GENCOM CONTROL PROG1 PROG2 Modifies the header of a program file called CONTROL.COM such that RSX files PROG1 and PROG2 become attached.

92 Operating systems

GET	GET temporarily gets console input from a specified disk file (rather than from the keyboard). Various options are provided to enable or disable echo to the console and to accept system input from a disk file. The command can also be used to restore input from the console.
HELP	The HELP command displays information on CP/M commands. HELP presents general information on a command as a topic and further detailed information as a subtopic. HELP can also be used with a command tail which directs the utility to a specified topic and subtopic rather than make use of the HELP menu list. By making additions to the HELP.DAT file, the HELP utility may also be customized to the user's own requirements. *Examples* HELP Produces a list of topics for which help is available. The user is then able to select the desired topic or exit. HELP ED COMMANDS Directs the HELP utility to provide information within the ED topic and COMMANDS subtopic.
HEXCOM	The HEXCOM utility takes a hexadecimal (HEX) file produced by the CP/M assembler utility and produces an executable (COM) program file. *Example* HEXCOM PROG1 Generates a program file, PROG1.COM, from a hexadecimal file, PROG1.HEX.
INITDIR	INITDIR is used to initialize a directory on the specified drive so that it permits permit time and date stamping.
LIB	The LIB utility creates and maintains library files that contain relocatable (REL) object code modules (see Chapter 5 for information on using a library utility).
LINK	The LINK utility combines relocatable object code modules into a program file with a COM extension. References to modules contained within the library are resolved by the LINK utility and appropriate code is included in the output file (see Chapter 5 for information on using a link utility).
MAC	The CP/M macro-assembler (MAC), is a utility which takes an assembly language source code file and assembles it to a standard Intel format hexadecimal object code file (see Chapter 5 for details on using a macro-assembler utility).
PIP	The PIP utility provides the user with a wide range of options involved with the copying and concatenation of files. PIP also allows the user to copy files to and from auxiliary devices. *Examples* PIP B:=A:PROG1.ASM Copies a file called PROG1.ASM from the disk in drive A to the disk in drive B. PIP B:PROG2.ASM=A:PROG1.ASM Copies a file called PROG1.ASM from the disk in drive A to the disk in drive B. The resulting file on drive B is given the name, PROG2.ASM.

	PIP B:=A:*.* Copies all files (with the exception of system files) from the disk in drive A to the disk in drive B. Where necessary, the COPYSYS utility should be used to transfer system files. PIP LST:=CON: Directs console input (i.e., the keyboard) to the list device (i.e., printer). Input should be terminated by CTRL-Z. PIP LST:=B:PROG1.ASM Sends the file named PROG1.ASM present on the disk in drive B to the list device (printer). PIP PRN:=B:PROG1.ASM Sends the file named PROG1.ASM present on the disk in drive B to the printer. This command differs from the previous command in that line numbers are included, tabs are expanded to every eighth column, and page ejects are produced after every sixtieth line.
PUT	Temporarily directs printer or console output to a disk file. PUT allows various options including enabling or disabling an echo facility and filtering control characters. *Example* PUT PRINTER OUTPUT TO FILE B:OUT.PRN Redirects all subsequent printer output to a disk file on drive B named OUT.PRN. The facility can be terminated by a command of the form PUT PRINTER OUTPUT TO PRINTER.
RMAC	The RMAC utility is a relocatable macro-assembler that takes assembly language source code (ASM) file and assembles them into relocatable (REL) program modules that can be linked to create executable program (COM) files (see Chapter 5 for details on using a macro-assembler utility).
SAVE	The SAVE utility is used to copy the contents of a specified block of memory to a file. SAVE is generally used in conjunction with a debugging utility (such as SID or ZSID) which is capable of loading files into memory. The SAVE utility prompts the user for a file specification as well as start and end addresses (in hexadecimal) of the memory block to be saved.
SET	The SET utility is used to select various file and drive attributes (including disk labels, date/time stamping and password protection). SET provides a range of options including the ability to mark a file as 'read only'. *Examples* SET B:[RO] Sets drive B to read-only status. Thereafter, the ERASE, PIP and RENAME commands will not function on any files present on a disk placed in drive B. The command may be disabled by entering CTRL-C in response to the system prompt, or by issuing a command of the form SET B:[RW]. SET B:*.COM[SYS,RO,PASS=F91X,PROT=READ] Makes all COM files present on the disk in drive B read-only system files.

94 *Operating systems*

SETDEF	The SETDEF utility is used to specify a number of system options including the order in which drives are searched. SETDEF allows the user to enable the system display mode and control page scrolling. *Example* SETDEF[NO PAGE] Turns off the default system page mode in which utilities pause after displaying a full console page. After issuing the command, the display will continue to scroll without waiting for the user to press a key.
SHOW	The SHOW command may be used to display various disk and drive related information. *Example* SHOW Displays the read/write status of all drives in a system and also specifies the remaining disk space.
SID	The SID (or ZSID) command invokes the symbolic debugger utility (see Chapter 5 for further details).

SUBMIT	The SUBMIT utility executes a sequence of commands contained in a batch file having a SUB extension. The batch file may be created using an editor (e.g. ED) and may contain commands and parameters required by CP/M utility programs. Up to nine arguments can be passed to batch files and these are marked $1 to $9.
	Examples
	Assume that the following batch file has been created and stored as STARTUP.SUB on the disk in the current default drive:
	
	SETDEF[DISPLAY]
	SET A:[RO]
	DATE SET
	
	   Thereafter, typing the command SUBMIT STARTUP will execute the contents of the STARTUP batch file which enables system display mode, sets drive A to read-only, and then prompts the user for the current date and time.
	   The following example shows how arguments can be passed using the SUBMIT command. Assume that the following batch file has been created and stored as PRINT.SUB on the disk in the current default drive PIP PRN:= $1:$2.TXT.
	   The command SUBMIT PRINT B HELP sends the file named HELP.TXT present on the disk in drive B to the printer. When SUBMIT command is executed, occurrences of $1 and $2 in the batch file are replaced by B and HELP, respectively.
	   The command SUBMIT PRINT A INFO will send the file named INFO.TXT present on the disk in drive A to the printer. In this case, occurrences of $1 and $2 in the batch file are replaced by A and INFO, respectively.

CP/M-Plus commands

SYSCOPY   The SYSCOPY command creates a new boot disk by copying the CPMLDR program. The SYSCOPY utility is somewhat more versatile than the COPYSYS utility and is able to support a wider range of disk formats. It is important to note that a cold start can only be performed by a disk which contains both the loader program (CPMLDR.COM) and the system file, CPM3.SYS. This file should be copied separately using the COPY or PIP utilities.
Example
SYSCOPY A: B: Copies CPMLDR.COM from the disk in drive A to the disk in drive B.

XREF   The XREF utility produces a cross-reference list of variables used in an assembly language program. Output may be directed to a printer using the $P option. The XREF utility requires access to the PRN and SYM files produced by the macro-assembler. These should be given the same filename as that specified in the XREF command tail.
Example
XREF B:PROG1 $P   Produces a cross-reference (XRF) file for PROG1 (taken from the disk present in drive B). Output is directed to the printer.

File specifications

Many of the CP/M commands make explicit reference to files. A file is simply a collection of related information stored on a disk. Program files comprise a series of instructions to be executed by the processor whereas data files simply contain a collection of records. A complete file specification has four distinct parts: a drive specifier, a filename, a filetype, and a password. The drive specifier is a single letter in the range A to P. The drive specifier is followed by a colon which separates it from the filename which follows. The filename comprises 1 to 8 characters while the filetype is usually specified in a 1 to 3 character extension. The filetype extension is separated from the filename by means of a full stop. When banked CP/M is used, a password may be added and this may consist of between 1 and 8 characters separated from the extension by means of a semi-colon. A complete file specification (filespec) thus takes the form:

[drive specifier]:[filename].[filetype];[password]

The filetype extension provides a convenient mechanism for distinguishing different types of files and CP/M-Plus provides various methods for manipulating groups of files having the same filetype extension. We could, for example, delete all of the assembly language source code files present on a disk using a single command of the form:

ERA B:*.ASM

Alternatively, we could copy all of the Intel hexadecimal format files from one disk to another using a command of the form:

PIP B:=A:*.HEX

Some commonly used CP/M filetype extensions are shown in Table 18.

The system control block

The system control block (SCB) is a CP/M-Plus data structure used primarily for communication between the BDOS and BIOS. The SCB also contains flags and data used by the CCP. The SCB is located within the BDOS but is also accessible to application programs.

Table 18 Common CP/M file types

Extension	Type of file
ASM	Assembly language source file.
BAK	Back-up file created by a text editor. The editor renames the source file with this extension and the revised file assumes the original file specification.
BAS	BASIC program source file.
COM	System executable program file.
HEX	Program file in hexadecimal format.
INT	CBASIC intermediate language file.
IRL	Indexed REL file produced by LIB.
LIB	Used by MAC and RMAC for macro libraries.
OVL	Program overlay file.
PRL	Page relocatable file (i.e., a file which can be relocated on any page boundary).
PRN	Printable file (displayable on a console or printer).
REL	Relocatable file produced by a macro-assembler.
SPR	System page relocatable file (i.e., system files required to generate CP/M-Plus).
SUB	Submit file (i.e., a file containing one or more CP/M-Plus commands).
SYM	Symbol table file.
SYS	System file.
TEX	Text file.
XRF	Cross-reference file.
$$$	Temporary file.

The high-order byte (FEH) of the various SCB addresses is defined within the SCB.ASM file. The SCB is page relocatable and GENCPM recognizes page relocatable addresses having a high-order byte of FEH as references to the SCB. Thereafter, when relocating the system, GENCPM changes these addresses to point to the SCB.

Table 19 shows the contents of the SCB together with offsets. Programmers wishing to make use of the SCB should be aware that the SCB contains certain critical system parameters. If a program attempts to modify such parameters, the outcome can be unpredictable and, at worst, will cause the operating system to crash. Furthermore, since the SCB is specific to CP/M-Plus (or CP/M 3) programmers should bear in mind the difficulty of porting programs to other CP/M environments (such as MP/M). These caveats excepted, access to the SCB can be vital to programmers wishing to develop system orientated applications. This theme is developed further under the heading 'Modifying the system control block'.

BIOS functions

The CP/M basic input/output system (BIOS) is a hardware dependent module that defines the low-level interface to a particular computer system and contains the requisite device drivers for peripheral I/O. Applications programmers rarely need direct access to the primitive functions within the BIOS and make use of the BDOS instead.

Table 19 System control block – fields and definitions

Offset	Definition
00–04	Reserved for system use (read only).
05	BDOS version number (read only).
06–09	Reserved for the user's own data or flags (read/write).
0A–0F	Reserved for system use (read only).
10–11	Program error return code (read/write). Used by a program to pass an error code or value to a chained program. CP/M-Plus uses this field to determine if a program executes successfully. BDOS function 108 (get/set program return code) is used to get/set this value.
12–19	Reserved for system use (read only).
1A	Console width (read/write). Contains the number of columns (characters per line) displayed on the console relative to zero. The value can be set using the DEVICE command or GENCPM utility. Most systems default console width to 79.
1B	Console column position (read only). Contains the current console column position.
1C	Console page length (read/write). Contains the number of lines of text displayed on the console. The value can be set using the

	DEVICE command or GENCPM utility. Most systems default console page length to 24.
1D–21	Reserved for system use (read only).
22–23	CONIN redirection flags (read/write). Each of the 16 bits represents a physical device. Bit 15 corresponds to device zero, bit 14 to device 1, and so on. Bits 0 to 3 (inclusive) are reserved for system use.
24–25	CONOUT redirection flags (read/write). Each of the 16 bits represents a physical device. Bit 15 corresponds to device zero, bit 14 to device 1, and so on. Bits 0 to 3 (inclusive) are reserved for system use.
26–27	AUXIN redirection flags (read/write). Each of the 16 bits represents a physical device. Bit 15 corresponds to device zero, bit 14 to device 1, and so on. Bits 0 to 3 (inclusive) are reserved for system use.
28–29	AUXOUT redirection flags (read/write). Each of the 16 bits represents a physical device. Bit 15 corresponds to device zero, bit 14 to device 1, and so on. Bits 0 to 3 (inclusive) are reserved for system use.
2A–2B	LSTOUT redirection flags (read/write). Each of the 16 bits represents a physical device. Bit 15 corresponds to device zero, bit 14 to device 1, and so on. Bits 0 to 3 (inclusive) are reserved for system use.
2C	Page mode (read/write). Used by some CP/M-Plus utilities to control scrolling. If page mode is not set to zero, the system scrolls data on the console without stopping.
2D	Reserved for system use (read only).
2E	Determines whether CTRL-H is interpreted as a RUBOUT/DELETE character (read/write). When set to zero, CTRL-H performs a destructive backspace (i.e., the cursor moves back and deletes). If set to FFH, CTRL-H is interpreted as a RUBOUT/DELETE character (deleted character is echoed).
2F	Determines if RUBOUT/DELETE is interpreted as a CTRL-H character (read/write). When set to zero, then RUBOUT/DELETE echoes the deleted character. If set to FFH, RUBOUT/DELETE is interpreted as a CTRL-H character (cursor moves back and deletes).
30–32	Reserved for system use (read only).
33–34	Console mode (read/write). Determines the action of certain BDOS Console I/O functions.
35–36	Reserved for system use (read only).
37	Output delimiter character (read/write). The default output delimiter character is $ but this can be changed using BDOS Function 110, get/set output delimiter.
38	List output flag (read/write). This byte controls the echo to the list device. When set to 1, console output is echoed to the list device. When set to 0, console output is not echoed to the list device.
39–3B	Reserved for system use (read only).
3C–3D	Current DMA address (read only). This value is initialized to 0080H by the CCP and also by BDOS function 13 (reset disk system). The DMA address can be modified by means of BDOS function 26 (set DMA address).
3E	Current disk (read only). This byte contains the currently selected

	default disk number. The value ranges from 0 to 15 (corresponding to drives A to P respectively). The BDOS function 25 (return current disk) can be used to determine the current disk value.
3F–43	Reserved for system use (read only).
44	Current user number (read only). This byte contains the current user number (in the range 0 to 15). BDOS function 32 (set/get user code) can modify or return the current user number.
45–49	Reserved for system use (read only).
4A	Multisector count (read/write). This byte is set by BDOS function 44 (set multisector count).
4B	Error mode (read/write). This byte is set by BDOS function 45 (set BDOS error mode). When the error mode byte is set to FFH, the system returns to the current program without displaying any error messages. When the error mode byte is set to FEH, the system displays error messages before returning to the current program. Any other value causes the system to terminate program execution and display error messages. BDOS function 45 (set BDOS error mode) can be used to modify this byte.
4C–4F	Drive search chain (read/write). These four bytes contain the numbers of the drives to be searched. The first drive to be searched has its number stored at offset 4C, the second at 4D, and so on. If less than four drives are to be searched, the next byte is set to FFH to indicate the end of the search chain. A drive number of 0 corresponds to the default drive while numbers 1 to 16 correspond to drives A to P respectively. The drive search chain can be displayed or set using the SETDEF utility.
50	Temporary file drive (read/write). This byte contains the drive number assigned to the temporary file drive. A drive number of 0 corresponds to the default drive whilst numbers 1 to 16 correspond to drives A to P respectively. The drive search chain can be displayed.
51	Error drive (read only). This byte contains the drive number of the selected drive when the last physical or extended error occurred.
52–56	Reserved for system use (read only).
57	BDOS flags (read only).
58–59	Date (read/write). The date is coded in natural binary and gives the number of elapsed days since 1 January 1978.
5A	Hours (read/write) in BCD format.
5B	Minutes (read/write) in BCD format.
5C	Seconds (read/write) in BCD format.
5D–5E	Common memory base address (read only). This value is zero for non-banked systems.
5F–63	Reserved for system use (read only).

BDOS functions

The basic disk operating system (BDOS) is the logical nucleus of CP/M and it provides a standard operating environment for transient programs and allows programmers access to its facilities by means of a number of function calls including those which deal with serial I/O (involving console, list and auxiliary devices) and the creation and maintenance of disk files.

The BDOS implements calls to the BIOS in order to perform hardware-dependent primitive functions. The software interface between BDOS and BIOS thus consists of a series of entry points into the BIOS. It should be stressed that the operating system's hardware independence stems from the fact that the BDOS routines remain constant (while the BIOS primitives are customized for a particular environment). The CP/M-Plus BDOS functions, together with entry parameters and returned values, are summarized in Table 20.

Making BDOS calls

A transient program may make BDOS calls in order to carry out a number of useful tasks and to alter certain operating system parameters. Programs must, however, obey the following standard conventions when making such calls:

- Prior to entering the BDOS, the C register must be loaded with the BDOS function and the DE register pair must be loaded with any byte, word, or address required.
- BDOS functions return single-byte values in register A and double-byte values in the HL register pair. BDOS functions also return with register A equal to register L and register H equal to register B.
- Where a transient program makes a call to BDOS using an unsupported function number, a value of FFFFH is returned in the HL register pair.

It is important to note that the BDOS does not restore registers to their entry values before returning to the calling program. It thus remains the responsibility of the programmer to ensure that all critical register values are saved prior to making the call and restored upon return.

The entry point for BDOS (in page zero) is address, 0005H. Hence calls within a transient program should reference this address. The neatest way of doing this is to use an EQU compiler directive of the form:

```
           BDOS    EQU   0005H      ; BDOS entry point
```

Individual BDOS functions can also be defined by EQU compiler directives as shown in the following example:

```
           CONIN   EQU   1          ; BDOS console input
           CONOUT  EQU   2          ; BDOS console output
```

The following example program shows how a simple call to BDOS function 1 (console input) can be made within a program which reads characters from the console until the RETURN or ENTER key is pressed:

```
; Various equates to aid readability
BDOS        EQU     0005H       ; BDOS entry point
CONIN       EQU     1           ; BDOS Console Input
RETURN      EQU     0DH         ; RETURN character
; The program itself
            CSEG                ; Assemble in Relative Mode
LOOPC:      MVI     C,CONIN     ; Set up function number
            CALL    BDOS        ; and make the call
            CPI     RETURN      ; Is it RETURN ?
            JNZ     LOOPC       ; No - so get more input
            RET                 ; Yes - so end the program
            END
```

Unlike most BDOS functions, BDOS console input does not require any parameter to be passed on entry (other than the function number which must be placed in the C register). When a parameter has to be passed from the calling program, this must be placed in the E register or DE register pair.

The following example written in 8080 assembly language sends control codes to an Epson-compatible printer so that it produces emphasized print. BDOS function 5 (list output) is used and this requires that the E register is loaded with the ASCII character to be sent to the list device prior to making the call.

```
; Various equates to aid readability
BDOS        EQU     0005H       ; BDOS entry point
LSTOUT      EQU     5           ; BDOS List Output
ESCAPE      EQU     1BH         ; ESCAPE character
BOLD        EQU     'E'         ; Emphasized control code
; The program itself
            ORG     100H        ; Base of TPA
BTYPE:      MVI     C,LSTOUT    ; Set up function number,
            MVI     E,ESCAPE    ; indicate that a control code follows,
            CALL    BDOS        ; and make the call.
            MVI     C,LSTOUT    ; Set up function number,
            MVI     E,BOLD      ; with code for bold type,
            CALL    BDOS        ; and make the call.
            RET                 ; End the program.
            END
```

Table 20 Summary of BDOS function calls

Function no:	name	Entry parameters	Returned values
0	System reset	C: 00H	None
1	Console input	C: 01H	A: ASCII char.
2	Console output	C: 02H A: ASCII char.	None
3	Aux. input	C: 03H	A: ASCII char.
4	Aux. output	C: 04H E: ASCII char.	None

102 Operating systems

5	List output	C: 05H E: ASCII char.	None
6	Direct console I/O	C: 06H E: FFH (input/status) or FEH (status) or FDH (input) or char. (output)	A: char. or status
7	Aux. input status	C: 07H	A: status (FFH = ready, OOH = not ready)
8	Aux. output status	C: 08H	A: status (FFH = ready, OOH = not ready)
9	Print using	C: 09H DE: string address	
10	Read console buffer	C: 0AH DE: buffer address	Console characters in buffer
11	Get console status	C: 0BH	A: Status (01H = default state, 00H = not ready)
12	Return version number	C: 0CH	HL: Version number
13	Reset disk system	C: 0DH	None
14	Select disk	C: 0EH E: Selected disk	A: Error flag H: Physical error
15	Open file	C: 0FH DE: FCB address	A: Directory code H: Physical or Extended error
16	Close file	C: 10H DE: FCB address	A: Directory code H: Physical or Extended error
17	Search for first	C: 11H DE: FCB address	A: Directory code H: Physical error
18	Search for next	C: 12H DE: FCB address	A: Directory code H: Physical error
19	Delete file	C: 13H DE: FCB address	A: Directory code H: Physical or extended error
20	Read sequential	C: 14H DE: FCB address	A: Error code H: Physical error
21	Write sequential	C: 15H DE: FCB address	A: Error code H: Physical error
22	Make file	C: 15H DE: FCB address	A: Directory code H: Physical or extended error
23	Rename file	C: 17H DE: FCB address	A: Directory code H: Physical or extended error
24	Return login vector	C: 18H	HL: Login vector
25	Return current disk	C: 19H	A: Current disk

26	Set DMA address	C: 1AH DE: DMA address	None
27	Get Addr (alloc) vector	C: 1BH	HL: ALLOC address
28	Write protect disk	C: 1CH	None
29	Get read-only vector	C: 1DH	HL: Read-only vector
30	Set file attributes	C: 1EH DE: FCB address	A: Directory code H: Physical or extended error
31	Get addr (DPB parms)	C: 1FH	HL: DPB address
32	Set/get user code	C: 20H E: FFH (get) or User code (set)	A: Current code or no value
33	Read random	C: 21H DE: FCB address	A: Error code H: Physical error
34	Write random	C: 22H DE: FCB address	A: Error code H: Physical error
35	Compute field size	C: 23H DE: FCB address	A: Error flag H: Physical or extended error (Random record field set)
36	Set random record	C: 24H DE: FCB address	Random record field set
37	Reset drive	C: 25H DE: Drive vector	A: 00H
40	Write random with zero fill	C: 28H DE: FCB address	A: Error code H: Physical error
41	Test and write record	C: 29H DE: FCB address	A: Error code H: Physical error
42	Lock record	C: 24H DE: FCB address	A: 00H
43	Unlock record	C: 2BH DE: FCB address	A: 00H
44	Set multisector count	C: 2CH E: Number of sectors	A: Return code
45	Set BDOS error mode	C: 2DH E: BDOS error mode	None
46	Get disk free space	C: 2EH E: Drive	A: Error flag H: Physical error First 3 bytes of current DMA buffer
47	Chain to program	C: 2FH E: Chain flag	None
48	Flush buffers	C: 30H E: Purge flag	A: Error flag H: Physical error
49	Get/set system control block	C: 31H DE: SCB Parameter block address	A: Returned byte HL: Returned byte
50	Direct BIOS calls	C: 32H	BIOS RETURN

104 Operating systems

59	Load overlay	DE: BIOS parameter block address C: 3BH DE: FCB address	A: Error code H: Physical error
60	Call resident system extension	C: 3CH DE: RSX parameter block address	A: Error code H: Physical error
98	Free blocks	C: 62H	A: Error flag H: Physical error
99	Truncate file	C: 63H DE: FCB address	A: Directory code H: Extended or physical error
100	Set directory label	C: 64H DE: FCB address	A: Directory code H: Physical or extended error
101	Return directory label data	C: 65H E: Drive	A: Director label data byte H: Physical error
102	Read file date stamps and password mode	C: 66H DE: FCB address	A: Directory code H: Physical error
103	Write file XFCB	C: 67H DE: FCB address	A: Directory code H: Physical error
104	Set date and time	C: 68H DE: DAT address	None
105	Get date and time	C: 69H DE: DAT address	A: Seconds DAT set
106	Set default password	C: 6AH DE: Password address	None
107	Return serial number	C: 6BH DE: Serial number field	Serial number field set
108	Get/set program return code	C: 6CH DE: FFFFH (get or program return code (set)	HL: Program return code (get)
109	Get/set console mode	C: 6DH DE: FFFFH (get) or console mode (set)	HL: Console mode (get)
110	Get/set output delimiter	C: 6EH DE: FFFFH (get) or E: Output delimiter (set)	A: Output delimiter (get)
111	Print block	C: 6FH DE: CCB address	None
112	List block	C: 70H DE: CCB address	None
152	Parse filename	C: 98H DE: PFCB address	HL: Return code Parsed file control block

Modifying the system control block

BDOS function 49 (get/set system control block) can be used to interrogate and/or modify parameters contained within the CP/M-Plus system control block (SCB). Function 49 requires the address of a 4-byte parameter block to be passed in the DE register pair. The first byte of the parameter block is used to specify the required SCB offset. The second byte of the parameter block indicates whether a byte or word is to be set or whether a get operation is to be performed on the following basis:

OOH	Indicates a get operation.
O1H–FDH	Reserved.
FEH	Indicates that a word is to be set.
FFH	Indicates that a byte is to be set.

When a get operation is being performed, the last two bytes within the SCB parameter block contain the byte or word to be set. This value will be written to the appropriate offset within the SCB. In calls which set bytes (rather than words), only the first byte of the two byte word is used.

When a get operation is being performed, a returned byte is placed in the A register while a returned word is deposited in the HL register pair.

The following example program modifies the action of the CTRL-H key combination so that it acts as a rubout key rather than a backspace key. The required SCB offset to modify CTRL-H action is 2EH and the byte value to be placed at this offset for rubout operation is FFH (see Table 19).

```
; Various equates to aid readability
BDOS      EQU   0005H       ; BDOS entry point
GSSCB     EQU   49          ; BDOS Get/Set SCB
CTRLH     EQU   2EH         ; Offset for CTRL-H control byte
BSET      EQU   0FFH        ; Set a byte in the SCB
WSET      EQU   0FEH        ; Set a word in the SCB
GSET      EQU   0           ; Get a value from the SCB
RUB       EQU   0FFH        ; Rubout delete
BSPACE    EQU   0           ; Backspace delete
OFFSET    EQU   SCBPB+0     ; Offset into the SCB
PSET      EQU   SCBPB+1     ; Indicates set/get and byte/word
VALUE     EQU   SCBPB+2     ; Byte or word to be set
; The program itself
          CSEG              ; Assemble in Relative Mode
RUBOUT:   MVI   A,CTRLH     ; Initialise parameter block
          STA   OFFSET      ; with CTRL-H SCB offset,
          MVI   A,BSET      ; indicate that this is a
          STA   PSET        ; byte operation and
          MVI   A,RUB       ; signal that rubout
          STA   VALUE       ; action is required.
          MVI   C,GSSCB     ; Set up function number,
          LXI   D,SCBPB     ; pass address of parameter block,
          CALL  BDOS        ; and (finally) make the call.
          RET               ; End the program.
; Parameter block for BDOS Function 49 calls
          DSEG              ; Data storage
SCBPB:    DS    4           ; Four byte parameter block
          END
```

Generating a new system

Finally, it may sometimes be necessary to modify the operating system in order to cater for a different hardware configuration. The GENCPM utility provides the means to reconfigure the system. The program prompts the user for new parameters, resolves references between the system modules, and creates a memory image file containing the BDOS and customized BIOS.

GENCPM can operate in automatic or manual display modes. In the former case, the utility requires access to a data file (GENCPM:DAT) which must be resident on the current drive. The prompts generated by GENCPM include default entries shown in parenthesis. The user is initially asked whether the GENCPM.DAT file should be used for defaults and whether a new GENCPM.DAT file should be created (see Figure 42). Further questions deal with console operation, the default drive, memory configuration, and disk parameters.

```
B>GENCPM

CP/M 3.0 System Generation
Copyright (C) 1982, Digital Research

Default entries are shown in (parens).
Default base is Hex, precede entry with # for decimal

Use GENCPM.DAT for defaults (Y) ?

Create a new GENCPM.DAT file (N) ?

Display Load Map at Cold Boot (N) ?

Number of console columns (#80) ?
Number of lines in console page (#24) ?
Backspace echoes erased character (N) ?
Rubout echoes erased character (N) ?

Initial default drive (A:) ?

Top page of memory (FB) ?
Bank switched memory (N) ?

Double allocation vectors (Y) ?

Accept new system definition (Y) ?
```

Figure 42 *Initial questions posed by the GENCPM utility program*

5
Assembly language programming

This chapter provides readers with an overview of assembly language programming techniques and explores the architecture and instruction sets of two of the most popular microprocessors used in low-cost process control applications. Rather than provide comprehensive information from which viable programs can be produced, the aim has been that of providing readers with sufficient information to make an informed choice of processor type, to outline the advantages and disadvantages of assembly language programming, and to introduce techniques used in developing assembly language programs.

Readers wishing to develop their own assembly language programs will not only require complete documentation for the processor type adopted (including a comprehensive explanation of the microprocessor's instruction set) but will also require development software comprising, as a minimum, a macro-assembler, a linking loader, and a symbolic debugger. Readers should not underestimate the investment required (in terms of both time and money) required to successfully follow this route.

Advantages of assembly language

Assembly language programs offer a number of advantages when compared with higher level alternatives. The principal advantages are that the executable code produced by an assembler (and linking loader) will:

- Invariably be more compact than an equivalent program written in a higher level language.
- Invariably run faster than an equivalent program written in a higher level language.
- Not require the services of a resident interpreter or a compiler run-time system.

- Be able to offer the programmer unprecedented control over the hardware in the system.

It is this last advantage, in particular, which makes assembly language a prime contender for use in bus-based process control applications. No other programming language can hope to compete with assembly language where control of hardware is concerned. Indeed, an important requirement of high-level languages used in process control applications is that they can be interfaced with machine code modules designed to cope with problems arising from limitations of the language where I/O control is concerned.

Disadvantages of assembly language

Unfortunately, when compared with higher level languages, assembly language has a number of drawbacks, most notable of which are that:

- Programs require considerably more development time (including writing, assembling, linking/loading and debugging) than their equivalent written in a high-level language.
- Programs are not readily transportable between microprocessors from different families. In practice this means that the system designer is constrained to select one particular microprocessor type (or family) and develop code exclusively for this particular device. There are, however, a few notable exceptions to this rule. Most 8080 and 8085 machine code routines will, for example, run on the Z80 (the former's instruction sets being upwardly compatible with the Z80). Furthermore, while the 8080/8085 and 8086/8088 are not directly code compatible, some degree of software portability is possible by virtue of a number of instructions which are common to the two devices.
- Unless liberally commented, the action of an assembly language program is not obvious from merely reading the source text. Programs written in high-level language are usually easy to comprehend.
- The production of efficient assembly language programs requires a relatively high degree of proficiency on the part of the programmer. Such expertise can usually only be acquired as a result of practical experience aided by appropriate training.

General concepts

Different microprocessors have different internal architectures and, in particular, the provision of registers accessible to the programmer will vary from one microprocessor to another. Some microprocessors (for example, the 68000) offer a *clean* internal architecture in which nearly every register can be described as *general purpose*). Others, for example, the 6502, may have a limited number of different registers largely dedicated to specific functions. Such obvious differences in architecture is reflected in corresponding differences in the type and function of the software instructions provided for the programmer.

The situation is further compounded by the fact that the microprocessor manufacturers frequently adopt different terminology to refer to the same thing. The variety of names used to describe the register used to indicate the outcome of the last ALU operation (and the internal status of the microprocessor) is a case in point. This is variously referred to as a flag register, status register, condition code register and processor status word.

Developing assembly language programs

The process of developing an assembly language program depends on a number of factors including the hardware system used and the available software. As a minimum, the task normally involves the following steps:

1. Analysing the problem and producing a specification for both hardware and software.
2. Developing the overall structure of the program, defining the individual elements and modules within it, and identifying those which already exist (or can be easily modified or extended) within the programmer's existing library.
3. Coding each new module required using assembly language mnemonics, entering the text using an editor, and saving each source code module to disk using an appropriate filename.
4. Assembling each source code module (using an assembler) to produce an intermediate relocatable object code file.
5. Linking modules (including those taken from the user's library) in order to produce an executable program.
6. Testing, debugging, and documenting the final program prior to evaluation and/or acceptance testing by the end-user.

In practice, the development process is largely iterative and there may also be some considerable overlap between phases. In order to ensure that the target specification is met (within the constraints of time and budget) an ongoing appraisal is necessary in order to maximize resources in the areas for which there is most need.

Software tools

The following items of utility software (software tools) are required in the development process:

1. An ASCII text editor (for example, ED supplied with CP/M-Plus).
2. A macro-assembler (for example, M80 supplied with the Microsoft utility software package).
3. A linking loader (for example, L80 supplied with the Microsoft utility software package).

In addition, two further software tools may be found to be invaluable. These are:

4. A cross-referencing utility (for example, CREF80 supplied with the Microsoft utility software package).
5. A library manager (for example, LIB supplied with the Microsoft utility software package).

Note: M80, L80, CREF80 and LIB are the four principal components of the Microsoft utility software package.

We shall briefly explain the function of each of these items of utility software in the production of assembly language programs.

Editors

Editors allow users to create and manipulate text files. Such files can be thought of as a sequence of keystrokes saved to disk. An assembly language source code file is simply a text file written using assembly language mnemonics and containing appropriate assembler directives.

The CP/M-Plus editor (ED) is invoked using a command of the form:

ED input file specification, output file specification

The output file specification is optional (in which case the output file specification will be the same as the input file specification).

The CP/M-Plus editor allows the user to operate on a file contained within a memory buffer. In effect, the editor treats the file as a long sequence of characters grouped together in lines. A character pointer moves within the buffer and identifies the point at which changes or inserts will be made. Various editor commands are available, including those for appending a specified number of lines of text (nA), moving the character pointer a specified number of characters backwards or forwards, deleting a specified number of characters (nD), moving the character pointer a specified number of lines backwards or forwards, inserting characters (I), and saving the text file to disk (W).

Format of source code statements

When preparing source text using an editor, it is important to bear in mind the requirements of the assembler concerning the format of source code statements. In the case of most Z80 assemblers (and Microsoft's Macro-80 in particular), each line of source code is divided into four fields, as shown in the example below:

Symbol Operation Argument Comments
```
COUNT:     DS         2000H     ; Events stored here
```

The first entry in the line of code is known as a symbol. The symbols used in a program are subject to certain constraints imposed by the assembler but are chosen by the individual programmer. Labels are a particular form of symbol which are referred to by one or more statements within a program. During the assembly process, labels (wherever they appear in the program) are replaced by addresses.

Entries in the operation field may comprise an operation code (opcode), a pseudo-operation code (pseudo-op), an expression, or the name of a macro. Operation codes are those recognized by the microprocessor as part of its instruction set (for example, LD, MOV, ADD, etc.) whereas pseudo-ops are directives which are recognized by the assembler and are used to control some aspects of the assembly process. Typical pseudo-ops are DB (define byte), DS (define storage), DW (define word), ORG (origin/program start address), and IN-CLUDE. The latter (equivalent to the MACLIB) pseudo-op, instructs the assembler to search a named macro library file and to expand macro definitions in terms of this library.

The argument field may contain constants or expressions (such as 0DH, 42, 64*32, 512/16, 'A', 'z' - 'A') or the operands required by microprocessor operation codes (represented by numbers, characters, symbols are extended opcodes).

The comment field contains a line of text, added by the programmer, which is designed to clarify the action of the statement within the program as a whole.

In the example shown previously, the variable COUNT has been declared in the symbol column. The operation field contains a pseudo-op (assembler directive) which instructs the assembler to reserve a byte of storage at the address specified in the argument (2000H). Thereafter, any references to COUNT will be taken to mean the byte stored at address 2000H. The programmer has added a comment (following the obligatory semi-colon) which reminds him/her that he/she is using COUNT to store the number of events counted.

Not all source code lines involve entries in all four fields, as in the next example:

```
Symbol    Operation  Argument  Comments
LD        A,COUNT    ; Get number of events
```

Here, the symbol field is blank since the instruction does not form part of the start of a block of code. The operation, LD, is an opcode which instructs the microprocessor to perform a LOAD operation. The operand required by the instruction specifies the accumulator (A) and COUNT (contents of memory location 2000H) as the destination and source respectively of data in the LOAD operation. The programmer has again added a brief comment to clarify the action of the line.

It should be noted that any line of source code starting with a semicolon is ignored by the assembler and treated as a comment. This allows the programmer to include longer comments as well as program or module headers which provide lengthy information on the action of the statements which follow.

Macro-assemblers

A macro-assembler can usually produce code which is either absolute or relocatable. In the former case, the code contains specific (absolute) address references whereas, in the latter, the code can be resident at virtually any uncontended address. In most cases, relocatable format is chosen since this allows program modules to be readily interfaced together to form complete programs.

The macro facility allows the programmer to write blocks of often-used code and incorporate these in programs by referring to them by name. The blocks of code are each defined as a macro. Thereafter, the macro assembler expands the macro call by automatically assembling the block of instructions which it represents into the program. The macro call can also be used to pass parameters to the assembler for use during the macro expansion.

As an example of the use of macros, the macro defined in the following code can be used to exchange the contents of two registers passed to the macro as parameters REG1 and REG2:

```
;
;             MACRO TO EXCHANGE REGISTER CONTENTS
;
;             PARAMETERS PASSED:  REG1, REG2
;
;             REGISTERS AFFECTED: REG1, REG2
;
SWAP    MACRO   REG1,REG2   ; Specify registers to swap
        PUSH    REG1        ; Stack REG1 first,
        PUSH    REG2        ; then REG2
        POP     REG1        ; REG1 receives REG2
        POP     REG2        ; REG2 receives REG1
        ENDM
```

The following code fragment shows how the macro call should be made:

```
LD      A,(DE)
ADC     A,(HL)
SWAP    DE,HL           ; Call the macro
INC     HL
DEC     DE
```

The macro-assembler expands the call, replacing it with the code given in its definition. The code generated by the macro-assembler will thus be equivalent to:

```
LD      A,(DE)
ADC     A,(HL)
PUSH    DE
PUSH    HL
POP     DE
POP     HL
INC     HL
DEC     DE
```

A macro facility can be instrumental in making significant reductions in the size of source code modules. Furthermore, macros can be nested such that a macro definition can itself contain references to other macros which, in turn, can contain references to others. A notable disadvantage of using macros is that the resulting object code may contain a large number of identical sections of code and will also occupy more memory space than if an equivalent subroutine had been used. In practice, therefore, programmers should use macros with care since there may be occasions where subroutines would be more efficient even though they may not be quite so easy to implement. Note, however, that it is necessary to have access to an area of RAM in order to implement a subroutine. In low-cost, stand alone systems, RAM space may be at a premium (or, at worst, non-existent) and hence the programmer may be forced to use ROM-based code which is entirely reliant on macro expansions rather than subroutines. We shall return to the topic of macros later in this chapter.

As well as macros, most assemblers also support conditional assembly. This allows the programmer to specify conditions under which portions of the program are either assembled or not assembled. Conditional assembly allows the programmer to test for specific conditions (using statements such as IF . . . ELSE . . . ENDIF) and use the outcome to control the assembly process.

Assemblers generally make two passes through a source file. During the first pass, macro calls are expanded and a symbol table is generated. On the second pass, relocatable code is generated which can be saved in a disk file. Such files are, however, not directly executable and require the services of a linking loader in order to function as programs.

Invoking a macro-assembler

The Microsoft macro-assembler (M80) is invoked by commands of the form:

M80 or
M80 = filename

The latter command instructs M80 to assemble the specified source file and save the result in a relocatable object file. This file will have the same name as the source code file but will be given a REL extension).

M80 assumes a filename extension of MAC hence the command:

M80 = SCAN

assembles a file previously saved to disk with the name SCAN.MAC and saves the relocatable object code in a disk file called SCAN.REL.

M80 can also generate an optional list file (with a PRN extension), and additional switches can be used to control the format of listings as well as instructing the macro-assembler to accept Z80 opcodes.

Linking loaders

A linking loader has essentially two functions. The first (loading) is concerned with placing a file in memory and assigning absolute addresses to the code and data in place of the relative addresses which appear in the output file of a macro-assembler. The second function (linking) is concerned with resolving external references such that any module which directs program execution outside itself (by means of a CALL, an external symbol, or an 'include') will be linked to the module which contains the corresponding code.

The resulting loaded and linked code modules can then be saved to disk so that they can later be treated by the operating systems as executable programs.

The Microsoft linking loader is invoked by a command of the form:

L80

followed by a command line which specifies a number of filenames and switches which control the loading and linking process.

Files are loaded in the order in which they are named in the command line, beginning at a default address of 103H under CP/M-Plus. Note that files need not be presented in strict order of execution as L80 places a jump instruction at address 100H which jumps to the start address of the first instruction to be executed, regardless of its location in memory.

Cross-reference utilities

A cross-reference facility can be used to produce a specially assembled file which lists the locations of all inter-module references and the locations of their definitions. This file is known as a cross-reference listing and it can be invaluable when debugging a program.

The Microsoft utility software package cross-reference utility (CREF) is invoked by commands of the form:

CREF80 or
CREF80 = filename

The cross-reference listing will be given the specified filename (together with a PRN filename extension). The PRN file may be printed or sent to the console using standard operating system commands. Additionally, CREF supports the same output designations as used by Microsoft's macro-assembler and hence commands of the form:

CREF PRN: = filename and
CREF TTY: = filename

are permissible. In the first case, the cross-reference listing will only be sent to the printer (not to a disk file). In the second case, the listing will only be sent to the console output device.

Library managers

A library manager allows the programmer to create run-time libraries from assembly language programs. The programs collected by a library manager may be special modules created by the programmer or modules from an existing library. The library manager allows the production of specialized run-time libraries for a variety of applications (including high-level language compilers).

The value of building a library is that the routine needed within a program can be very easily linked into an executable object code file. Routines taken from the library can be used to construct further libraries or combined into relocatable modules. The Microsoft utility software package library manager utility (LIB) is invoked by the command:

LIB

Thereafter, library manager commands consist of an optional destination

field (filename), a source field (filenames), and an optional switch field.
The command:

TRLIB = SIN,COS,TAN,ATAN,ACOS/E

for example, generates a relocatable file, TRLIB.REL, concatenated from the following files: SIN.REL, COS.REL, TAN.REL, ATAN.REL and ACOS.REL.

Symbolic debuggers

A symbolic debugger is an item of utility software which facilitates interactive testing and debugging of programs. As a minimum, a debugger should provide the user with commands which can be used to:

- Examine and modify memory.
- Examine and modify CPU registers.
- Run a program (starting at a given address) with breakpoints at which execution may be halted to permit examination of the CPU registers
- Single-step a program (starting at a given address) with a register dump at the completion of each instruction.
- Disassemble a block of memory into assembly language mnemonics.
- Relocate a given block of memory.
- Initialize a given block of memory with specified data.
- Load/save blocks of memory from/to disk.

Utilities provided with CP/M-Plus include a symbolic interactive debugger (SID) designed for use with the 8080 microprocessor and ZSID, a similar debugging tool for use with the Z80 microprocessor. Both programs use the same set of single-letter commands (listed in Table 21) which may be used with one, or more, additional parameters.

Table 21

Letter	Command	Function
A	Assemble	Begins line-assembly at a specified address.
C	Call	Calls the code resident at a specified address (without disturbing the state of the CPU).
D	Display	Displays memory contents (in various formats) starting and ending at specified addresses (see Figure 43).
F	Fill	Fills memory with given data starting and ending at specified addresses.
G	Go	Runs the program from a program counter value with, or without, breakpoints.
H	Hex	Performs hexadecimal arithmetic and number conversion.

Symbolic debuggers 117

I	Input	Initializes default areas in low-memory prior to using the R command.
L	List	Provides a disassembled listing of a block of memory between given starting and ending addresses (see Figure 43).
M	Move	Moves a block of memory of given size from a given start address to a given destination address.
P	Pass	Monitors the passage of a program through a specified program counter value (the 'pass point') and allows permanent breakpoints to be set after passing through the point a specified number of times.
R	Read	Reads code and/or symbols in preparation for a program test.
S	Set	Sets memory starting at a specified address, and with automatic increment, in either byte or word format.
T	Trace	Traces a specified number of program steps, with or without calls.
U	Untrace	Performs a trace function but does not display register contents.
X	Examine	Displays the current state of the CPU registers.

```
#DC000
C000: F5 CD 9D C5 CD A5 C5 F1 C2 94 C3 C3 BE BF CD 5C  ...............\
C010: C3 CD 80 C6 07 07 07 07 E6 F0 F5 CD 5C C3 CD 80  ............\...
C020: C6 C1 B0 C9 C5 E5 D5 CD 0E C0 47 D1 82 57 78 E1  ..........G..Wx.
C030: C1 C9 21 00 B3 CD 93 C5 DA 42 C0 3A 85 CC B7 CC  ..!......B.:....
C040: 16 C4 3E 55 01 4C 54 CD 86 C5 F5 21 36 CD 11 5C  ..>U.LT....!6..\
C050: 00 0E 10 7E 12 23 13 0D C2 53 C0 AF 32 7C 00 3A  ...~.#...S..2|.:
C060: 5D 00 FE 20 CA F0 C0 21 86 CC CD D1 C5 CD 37 C5  ].. ...!......7.
C070: 3C CA 94 C3 CD 5C C3 FE 1A CA F0 C0 FE 21 DA 74  <....\.......!.t
C080: C0 CD 11 C0 F5 CD 0E C0 D1 5F 2A 7F CC 19 E5 CD  ........._*.....
C090: 5C C3 FE 20 CA A3 C0 E1 CD 5C C3 FE 20 DA 77 C0  \.. .....\.. .w.
C0A0: C3 98 C0 2A 06 00 1E 00 2B CD 5C C3 FE 09 CA C6  ...*....+.\.....
C0B0: C0 FE 0D CA C6 C0 FE 21 DA 94 C3 77 1C 7B FE 10  .......!...w.{..
#LC000
 C000    PUSH PSW
 C001    CALL C59D
 C004    CALL C5A5
 C007    POP  PSW
 C008    JNZ  C394
 C00B    JMP  BFBE
 C00E    CALL C35C
 C011    CALL C680
 C014    RLC
 C015    RLC
 C016    RLC
```

Figure 43 *Typical debugger display showing a hexadecimal and ASCII dump of a block of memory together with a disassembly of the code starting at C000H*

The Z80 microprocessor

The Z80 is a powerful 8-bit microprocessor which has a total of 158 basic instructions (including bit test and set, and block move). The Z80 was designed by Zilog as a very much enhanced 8080 device and yet retains full upward operation code compatibility with that device.

The Z80 has seventeen internal registers including a complete duplicate set of general purpose registers (known as the *alternate register set*). Two 16-bit index registers (IX and IY) are provided as well as a 16-bit stack pointer (SP) and 16-bit program counter (PC). The processor also provides three distinct interrupt modes including one which is directly equivalent to the 8080 interrupt response.

An interesting feature of the Z80 (and one which has earned it considerable popularity among equipment manufacturers) is that it incorporates provision for refreshing dynamic RAM.

For control applications, the Z80's instruction set contains IN and OUT instructions which permit reading from and writing to I/O ports. Memory and port addresses are distinguished by memory request (MREQ) and input/output request (IORQ) lines.

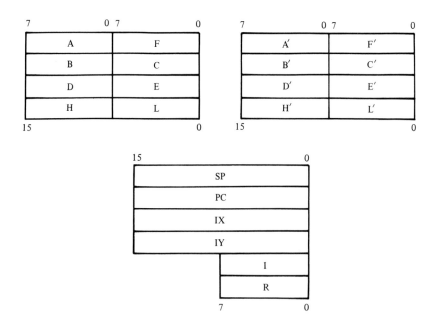

Figure 44 *Z80 register model*

Z80 registers

The register model for the Z80 is shown in Figure 44. The diagram has been divided into three distinct blocks showing dedicated registers, general purpose registers, and the alternate general purpose register set. We shall briefly examine the function of each register together with representative instructions.

The accumulator

The accumulator (A) is the primary source and destination for data used in a large number of single operand instructions. The accumulator also supplies one of the operands required by arithmetic and logical instructions, the results of which are returned to the accumulator.

As with all of the general purpose registers, the accumulator may be loaded with immediate data (i.e., the data is the operand) by means of instructions of the form LD register, data. For example, the instruction LD A,7FH loads the accumulator with 7FH. The result of executing this instruction is that, with the exception of bit 7, all bits within the accumulator will become set.

The occasion often arises when it is necessary to clear the accumulator (i.e., ensure that all bits in the accumulator are reset). This can be achieved by the 2-byte instruction LD A,00H. A neater method, however, makes use of the instruction XOR A (the result of exclusive-ORing a register with itself is always zero). The advantage of this method is simply that it only requires a single byte of machine code.

The accumulator may be loaded with data present at an address in memory by means of instructions of the form LD A, (address). The instruction LD A, (8001H), for example, will copy the byte present at address 8001H into the accumulator. The brackets around the address should be thought of as meaning 'contents of'. Hence the instruction may be written; 'load accumulator with the contents of hexadecimal address 8001'.

The reverse of such operations (i.e., copying a byte currently present in the accumulator to an address in memory) is performed by instructions of the form LD (address),A. In the case of the address previously used, the required instruction would be LD (8001H),A. The following code fragment copies the byte present at address 01FFH to address 0200H:

```
LD        A,(01FFH)    ; Get data from first address
LD        (0200H),A    ; and copy to the second
```

B and C registers

The B and C registers provide secondary data storage and, as such, can function as secondary accumulators, each 8-bits in length. The two registers can also be used 'end-on' to form a single 16-bit register. This register can be used as a data pointer (i.e., it can be used to hold an address at which data is to be stored or from which data is to be loaded) and also used to implement a count. In the latter case, the register pair is loaded with a 16-bit value and successively incremented or decremented until the value reaches zero (in which case the zero flag becomes set). The following code fragment shows how the BC register pair can be used to implement a typical 16-bit delay subroutine:

```
DELAY:     LD      BC,TIMES     ; Number of times round the loop.
BEGIN:     DEC     BC           ; Count down.
           LD      A,B          ; Has the count
           OR      C            ; reached zero ?
           JR      NZ,BEGIN     ; No, keep going.
           RET                  ; Yes, then exit.
```

A further application of the BC register pair, and one which is particularly relevant to control applications, is that of providing an extended port address. The normal IN A,(port) and OUT (port),A instructions possessed by the Z80 are able to reference any one of 256 different port addresses (specified by port). Where I/O is to be provided to addresses outside the range 00H to FFH (as is the case with the STE bus system which permits 4096 I/O locations), the instructions IN register,(C) and OUT (C),register can be used. These instructions perform in the same way as the normal IN and OUT instructions but the contents of the B register appear on the upper half of the address bus, thereby allowing for a total of 65536 different port addresses.

The following fragment of code outputs the value currently in the accumulator to port address FFFH and then immediately inputs the value from the port address given by INPORT to the E register:

```
LD      BC,0FFFH     ; Output port address in BC
OUT     (C),A        ; Output the byte
LD      BC,INPORT    ; Input port address in BC
IN      E,(C)        ; Input the byte
```

D and E registers

The D and E registers function in exactly the same manner as the BC register pair but with the exception of the extended I/O facility.

H and L registers

The H and L registers can also be used individually as secondary accumulators and jointly as a 16-bit register pair. The HL register is, however, often reserved for use as a pointer to an address used for data storage. Data can readily be transferred between any of the general purpose registers (including the accumulator) and the memory location addressed by the HL register pair.

Immediate data can be loaded into memory using the LD (HL),data instruction whereas data present in any of the general purpose registers can be copied to memory using the LD (HL), register instructions. Conversely, data present in memory can be copied to a register using the LD A,register. The address pointer formed by register pair HL can also be easily incremented or decremented by means of the INC HL and DEC HL instructions.

The following subroutine loads memory addresses 1000H to 10FFH with data values of 00H to FFH respectively:

```
LOADMEM:   LD    HL,1000H      ; Start at 1000H.
           XOR   A             ; The first data item is 0.
LOADIT:    LD    (HL),A        ; Copy data to memory.
           INC   HL            ; Increment the pointer,
           INC   A             ; and the data.
           CP    FFH           ; All 256 locations loaded?
           JR    NZ,LOADIT     ; No, keep going.
           RET                 ; Yes, then exit.
```

Flag register

The composition of the Z80's flag register (F) is shown in Figure 45. There are six flags, each of which may either be set or reset depending upon the internal status of the CPU. Flags keep their status (either set or reset) until an instruction is executed which has an effect on them. Their functions may be summarized in Table 22.

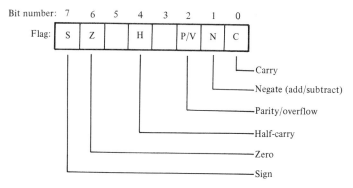

Figure 45 *Z80 flag register*

Alternate registers

The alternate register set performs in exactly the same manner as the main

Table 22

Flag	Function
Carry	Set whenever a carry bit is generated by an arithmetic operation. Reset by Boolean operations.
Negate	Set by all subtract operations. Rest by all add instructions.
Parity/overflow	Set when an overflow occurs during an arithmetic operation. Set (to indicate even parity) or reset (top indicate odd parity) by input, rotate and Boolean operations. Set during block transfer and search operations but reset when the byte counter reaches zero.
Auxiliary carry	Set when a carry is generated from bit 3 to bit 4 during an arithmetic operation, otherwise reset.
Zero	Set when an arithmetic or Boolean operation produces a zero result. Reset when a result is non-zero.
Sign	Set when the MSB of the result of an arithmetic or Boolean operation is also set, otherwise reset.

set of general purpose registers. Various exchange instructions are provided so that the contents of equivalent register pairs can be exchanged.

As an example, the instruction EX AF,AF' exchanges the contents of the AF and AF' register pairs whereas the instruction EXX exchanges the contents of the remaining general purpose registers with their equivalents in the alternate set.

Stack pointer

The stack pointer (SP) is a 16-bit register which allows the programmer to establish a stack for temporary storage of data and addresses. The stack pointer may be loaded with immediate data using LD SP,data or loaded with data from memory using LD SP,(address). A further alternative is that of moving the contents of the address pointer, HL, to the stack in which case the required instruction is LD SP,HL. The stack pointer may be incremented and decremented by means of the instructions INC SP and DEC SP, respectively. Note that the stack pointer is automatically incremented and decremented by the POP and PUSH instructions respectively.

The following code fragment can be used to save the contents of the main and alternate register sets on the stack prior to the execution of an interrupt service subroutine:

```
PUSH      AF              ; Save main registers
PUSH      BC
PUSH      DE
PUSH      HL
```

```
PUSH    IX              ; Save index registers
PUSH    IY
EX      AF,AF'          ; Swap to the alternate
EXX                     ; register set
PUSH    AF              ; Save alternate registers
PUSH    BC
PUSH    DE
PUSH    HL
```

To restore the register contents to their values before the service routine, the following code should be used (notice the order in which data is POPped off the stack is the reverse of that which was used to PUSH it into the stack):

```
POP     HL              ; Restore alternate registers
POP     DE
POP     BC
POP     AF
EX      AF,AF'          ; Swap to main
EXX                     ; register set
POP     IY              ; Restore index registers
POP     IX
POP     HL              ; Restore main registers
POP     DE
POP     BC
POP     AF
```

Index registers

The Z80's two index registers (IX and IY) provide the programmer with the ability to refer to memory locations in terms of an offset (specified in a data movement instruction) relative to a variable base address (held in the index register concerned).

Immediate data may be loaded into memory using instructions of the form LD (IX+displacement),data and LD (IY+displacement),data. The address at which the data is stored is found by adding the contents of the named index register to the specified displacement (in the range 0 to 255).

Instructions of the form LD register,(IX+displacement) and LD register,(IY+displacement) allow the programmer to load a specified register with data at an indexed address (again found by adding the contents of the named index register to the specified displacement), whereas the instructions LD (IX+displacement),register and LD (IY+displacement),register perform the reciprocal operation.

Interrupt vector

The interrupt vector (I) is used to store the most significant byte (i.e., page number) of a table of vectors (maximum 256) to individual service routines. The actual start address of each service routine is placed in the vector table and the particular entry in the table is found from the offset supplied by the interrupting device (mode 2). The start address of the required subroutine is thus stored at an address which itself is formed by multiplying the contents of the I register by 256 and adding the value of the byte passed by the interrupting device on the data bus. This byte of data is made available by the interrupting device in response to an interrupt acknowledgement generated by the processor.

Refresh register

The refresh register (R) is used to control the refreshing of dynamic RAM. The register is of little practical value to the programmer.

The 8086 microprocessor

The first of the second-generation 16-bit microprocessors to become available was Intel's 8086. This rapidly became an industry-standard device, featuring in applications ranging from personal computers to industrial process controllers. More importantly, the 8086 was designed with modular internal architecture, which has been the theme for a series of upwardly compatible microprocessors. This approach has also allowed Intel to produce a virtually identical microprocessor having an 8-bit external bus. This 8088 device permits the use of lower-cost external support devices and thus provides the equipment manufacturer with an excellent compromise between process power and system hardware cost.

Intel has also retained a degree of upward compatibility between the older 8080 and 8085 8-bit devices and the newer 8086 and 8088 processors. The 8080/8085 registers form a subset of the 8086/8088 registers and a number of 8080/8085 instructions have equivalents in the 8086/8088 instruction set. Furthermore, since the Z80 is upwardly compatible with the 8080, a number of common Z80 instructions also have equivalents in the 8086/8088 instruction set even though different mnemonics are employed.

The 8086 has fourteen 16-bit registers (including those which are the equivalents of the 8080/8085 register set). The CPU has twenty address lines and thus provides for a 1 megabyte address range. The I/O address range, on the other hand, is 64 kilobytes. The 8086 has a large instruction set, a total of twenty-five different addressing modes.

The register model of the 8086 is shown in Figure 46. Of the fourteen 16-bit registers available, four may be described as general purpose and can be divided into separate 8-bit registers. As an example, the 16-bit extended accumulator (AX) can be divided into two 8-bit registers, AH and AL. The high-byte of a 16-bit word placed in AX is stored in AH while the low-byte is stored in AL. Instructions can be made to refer to various parts of the accumulator so that operations can be carried out on the word stored in AX, or the individual bytes stored in AH or AL.

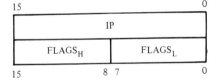

Figure 46 *8086 register model*

The four segment registers are code segment (CS), data segment (DS), stack segment (SS) and extra segment (ES). By making appropriate changes to the contents of these registers, the programmer can dynamically change the allocation of work space.

Readers may be forgiven for wondering how it is that a processor having only 16-bit internal registers is capable of generating 20-bit addresses. The 8086 forms a 20-bit address from the contents of one of the segment registers (either CS, DS, SS, or ES) and an offset taken from one (or more) of the other registers or from a memory reference within the program. The four segment registers (CS, DS, SS, and ES) effectively allow the programmer to set up individual 64 kilobyte work spaces within the total 1 megabyte address range.

The actual 20-bit physical memory address is formed by shifting the segment address four bits to the left (adding four least significant bits), which effectively multiplies the segment register contents by sixteen. The contents of the instruction pointer (IP), stack pointer (SP) or other 16-bit memory reference is then added to the result. This process is illustrated in Figure 47.

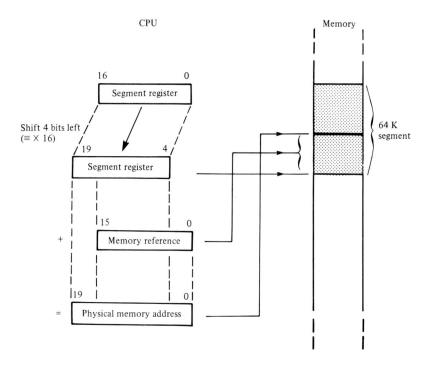

Figure 47 *Using a segment register to form a physical memory address*

8086 registers

Accumulator, AX (AH and AL)

As usual, the accumulator is the primary source and destination for data used in a large number instructions. The data movement instructions in Table 23 give some idea of the range of options available.

Table 23

MOVBI	AL,data	Moves 8-bit immediate data into the least-significant byte of the accumulator (AL). Equivalent to a Z80 instruction of the form LD A,data.
MOVBI	AH,data	Moves 8-bit immediate data into the most-significant byte of the accumulator (AH).
MOV	AX,register	Copies the contents of the specified register to the 16-bit extended accumulator (AX).
MOVB	AH,AL	Copies the byte present in register AL to register AH.

BX (BH and BL) register

The BX register is normally as a base register (address pointer). As such, it is equivalent to the HL register pair found in the Z80. The data movement instructions in Table 24 give some idea of the range of options that are available:

Table 24

MOVI	BX,data	Moves 16-bit immediate data into the general purpose base register (BX).
MOV	BX,address	Copies the 16-bit word at the specified address into the general purpose base register (BX). Equivalent to the Z80 instruction, LD HL,(address).
MOV	BX,register	Copies the contents of the specified register to the base register (BX).
MOV	[BX],AL	Copies the contents of the AL register to the memory address specified by the BX register.

CX (CH and CL) register

The CX register is normally used as a loop counter in much the same manner as the Z80 BC register pair. The 8086 LOOP instruction tests the contents of the CX register pair in order to determine whether the loop should be repeated or not. This makes coding loops extremely simple as the following code fragment shows:

```
DELAY:   MOVI   CX,0C00H    ; Number of times round the loop
         LOOP   DELAY       ; Finished ?
         RET
```

128   *Assembly language programming*

The CX and CL registers are also used to implement repeated string moves, shifts and rotates. The following example shows how the contents of the accumulator can be rotated by the value placed in the CL register:

```
ROTE4:   MOVBI    CL,4       ; Number of bits to
         ROR      AX,CL      ; Rotate to the right
         RET
```

DX (DH and DL) register

The DX register is a general purpose 16-bit register which can also be used as an extension of the AX register in 16-bit multiplication and division. The DX register is roughly equivalent to the Z80 DE register pair.

Stack pointer

The function of the 8086 stack pointer (SP) is similar to that of the same equivalent register in the Z80 and hence will not be discussed separately here.

Base pointer, destination index and source index

The base pointer (BP), destination index (DI) and source index (SI) registers are used in some of the more sophisticated of the 8086 addressing modes which permit the programmer to implement advanced data structures (such as two-dimensional arrays). All three registers are used to form addresses as shown in Table 25.

Table 25

MOV	[BP+20],AX	Copies the word present in the AX register to an address offset by 20 bytes from the base pointer (BP).
MOVBI	[DI],SPACE	Places 20H (previously defined by an equate of the form SPACE EQU 20H) at the address pointed to by the destination index (DI).
MOVI	SI,MESS1	Move the start address of MESS1 into the source index (SI). Thereafter, SI can be used with an offset to point to a particular character within the string, MESS1.

Instruction pointer

The function of the 8086 instruction pointer (IP) is similar to that of the Z80 program counter and hence will not be discussed separately here.

Flag register

The 8086 has a 16-bit flag register (F) which contains nine status bits which may be set or reset depending upon the internal state of the CPU. Flags keep their status (either set or reset) until an instruction is executed which has an effect on them. The 8086 flags are shown in Figure 48.

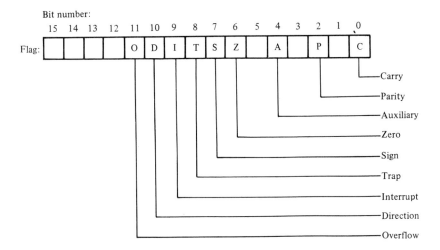

Figure 48 *8086 flag register*

Segment registers: CS, DS, SS, ES

We have already briefly mentioned the function of the four segment registers. Each register is associated with a separate workspace. The workspace defined by the code segment register will contain program instructions while the space defined by the data and extra segments will generally contain data. In situations where RAM is limited (i.e., when only a small proportion of the total 1 megabyte address range is populated), there is no reason why the several segment registers should not have the same value. The code fragment:

can be used to make the data segment equal to the code segment (note that the instruction MOV DS,CS is not valid).

As a further example, the code fragment:

```
MOV        AX,VIDRAM        ; Make Data Segment point
MOV        DS,AX            ; to Video RAM
```

can be used to make the data segment point to the start of a block of video RAM (VIDRAM will previously have been the subject of an equate).

Assembly language programming techniques

Regardless of the processor involved, a number of techniques can be used to improve the overall efficiency of a program. Many of these techniques are easy to implement and merely require forethought and self-discipline on the part of the programmer.

Structured programming

The overall structure of the program should be defined at a very early stage and no attempt should be made at coding any of the modules required by the program until the overall program structure has been finalized.

The structure of the program should be defined using one or more flowcharts or by means of a number of suitably structured statements in English. In any event, the overall flow of the program should be sequential and all transfers of control (i.e., jumps and calls) should be explicit.

As an example of using flowcharts and structured English statements, consider the case of a fragment of code which reads an input port, loops until a byte having its most significant bit set is detected, and then transfers this byte to an output port.

A flow chart for the process is shown in Figure 49. In addition (or perhaps as an alternative) to using a flowchart, we could express the problem in terms of the following structured English statements:

Begin
Repeat
   Get byte from PORT1
Until MSB of byte is set
Output byte to PORT2
End

Armed with one or other of the foregoing algorithms, it is a relatively simple matter to develop the code. A particular solution, based on the Z80 microprocessor, would take the following form:

Assembly language programming techniques   131

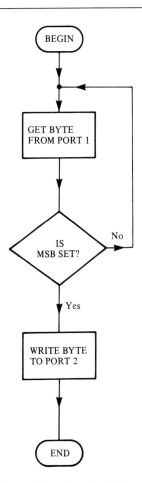

Figure 49 *Flowchart for a simple I/O process*

```
;                 READ PORT1, LOOP UNTIL MSB SET, THEN
;                 TRANSFER BYTE TO PORT2
;
;                 ENTRY: n/a
;
;                 EXIT: A = (PORT1), BC = PORT2, ZF set
;
;                 REGISTERS AFFECTED: A, B, C, F
;
GETBYTE:    LD    BC,PORT1         ; Get byte from
            IN    A,(C)            ; PORT1
            BIT   7,A              ; Is MSB set?
            JR    NZ,GETBYTE       ; No, keep trying
            LD    BC,PORT2         ; Yes, send byte
            OUT   (C),A            ; to PORT2
            RET
```

Subroutines

The code shown in the previous example constitutes a subroutine. It is called by name from various points in the main program (using the CALL instruction) and returned to (by means of a corresponding RETurn) instruction. If desired, both the CALL and RETURN instructions can be made conditional on the contents of the flag register. Furthermore, a subroutine may have several conditional RETurn statements.

The CALL instruction saves the old value of the program counter in the stack before replacing it with the value of the subroutine start address. On returning from the subroutine, the program counter is loaded with the value saved on the stack so that the main program can be resumed at the point at which it was left.

Parameters can be easily passed to subroutines by simply placing them in one or more of the CPU registers. Alternatively, parameters may be passed using the stack or by reserving an area of memory in which parameters can be deposited before making the call and recovered after the call has been made. These techniques allow the passing of a much greater number of parameters than would be possible using just the CPU registers.

Care must be taken to preserve the contents of any CPU registers that may be modified as a result of executing the subroutine call and that are required in subsequent processing. It is thus essential to have a knowledge of the effect of a subroutine on the CPU registers (in any event, this should be clearly indicated in the source code). Furthermore, subroutines should be designed so that they minimize usage of the CPU registers, thus keeping things simple for the programmer and reducing any potential overhead associated with storing and retrieving register contents.

Interrupt service routines

Interrupts may be generated by a range of support devices and external hardware and are used to alert the CPU to the fact that an external device needs attention. Furthermore, since each interrupting device will require its own customized service routine, some mechanism will be required for identifying the source of the interrupt and vectoring to the appropriate interupt service routine.

An interrupt service routine operates in a very similar manner to a subroutine. The service routine is stored away from the main body of code and effectively called when the interrupt occurs. Unlike a subroutine, however, there is no need for a CALL instruction. Since interrupts may occur at virtually any point in the execution of the main program (provided they have not been disabled or 'masked'), the response must be automatic; the CPU must suspend its current task and save the return address so that the program can be resumed at the point at which it was left. Thereafter, the programmer is responsible for preserving the state of any other registers which may be changed during execution of the interrupt service routine.

The Z80 has three different interrupt modes and the programmer can select a desired mode by means of the IM instruction. Interrupts can be enabled or disabled by means of the EI and DI instructions respectively.

Z80 interrupt mode 0 has been incorporated to retain compatibility with the interrupt mechanism implemented on the 8080 microprocessor. In mode 0, the CPU uses the data passed by an interrupting device to determine the address at which the interrupt service routine can be found. The data byte (one of eight possible vectors) is placed on the bus in response to an interrupt acknowledgement generated by the CPU and is directly equivalent to a single-byte restart (RST) instruction. The values of the data byte (together with its equivalent RST instruction and destination address) are given in Table 26.

Table 26

Data byte (opcode)	Equivalent RST instruction	Destination address
C7	RST 0H	0000H
CF	RST 8H	0008H
D7	RST 10H	0010H
DF	RST 18H	0018H
E7	RST 20H	0020H
EF	RST 28H	0028H
F7	RST 30H	0030H
FF	RST 38H	0038H

RST 0 is equivalent to a hardware CPU RESET and therefore is of little use. The programmer is thus able to implement seven different interrupt service routines in mode 0. Note that it is not necessary to have the complete service routine present in page zero since code can be easily vectored to another address by means of one or more JUMP (JP) instructions.

Mode 1 is rather more limited as it only offers a single interrupt service routine which should be located at 0038H or JUMPed to from this address. A mode 1 interrupt is equivalent to an RST 38H instruction and no data is passed by the interrupting device.

The last of the Z80's interrupt modes offers programmers a great deal of flexibility and allows for a maximum of 256 different interrupt service routines. When the interrupt request is accepted, the CPU disables further interrupts (so that the interrupt service routine cannot itself be interrupted) and saves the program counter in the stack. The CPU then reads the data passed on the bus by the interrupting device in response to the interrupt acknowledgement and uses this to form the address of a pointer to the required interrupt service routine (see interrupt vector for further details). Execution continues from the start address of the particular interrupt service routine. The first few lines of code within the interrupt service routine must be dedicated to preserving the contents of all CPU registers which may be affected by the routine. Register contents may be preserved in the stack (see information given earlier relating to the Z80 stack pointer) or may be preserved in the alternate register set.

After completing execution of the main body of the interrupt service routine, the routine restores any registers that were saved and then enables interrupts (using an EI instruction). The routine then closes with a RETI (RETurn from interrupt) instruction.

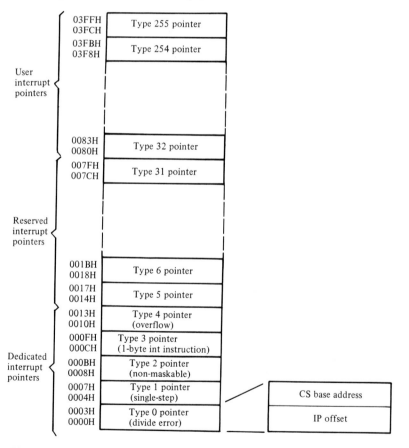

Figure 50 *8086 interrupt pointer table*

The 8086 provides somewhat superior interrupt handling and uses a table of 256 4-byte pointers stored in the bottom 1 K of memory (addresses 0000H to 03FFH). Each of the locations in the interrupt pointer table can be loaded with a pointer to a different interrupt service routine. Each pointer contains 2 bytes for loading into the code segment (CS) register and 2 bytes for loading into the instruction pointer (IP). This allows programmers to place their interrupt service routines at any suitable location within a megabyte of physical address space.

Each of the 256 interrupt pointers is allocated a different type number. A type 0 interrupt has its associated interrupt pointer in the lowest 4 bytes of memory (0000H to 0003H). A type 1 interrupt will have its pointer located in the next 4 bytes of memory (0004H to 0007H), and so on.

The structure of the 8086 Interrupt Pointer Table is shown in Figure 50. Interrupt types 0 to 4 have dedicated functions while types 5 to 31 are reserved. Hence there are 224 remaining locations in which interrupt pointers may be placed. As with the Z80 mode 2, the interrupting device places a byte on the data bus in response to an interrupt acknowledgement generated by the CPU. This byte gives the interrupt type and the 8086 loads its code segment and instruction pointer registers with the words stored at the appropriate locations in the interrupt pointer table and then commences execution of the interrupt service routine.

The following code fragment shows how the interrupt pointer table can be initialized:

```
; Initialise Interrupt Pointer Table
        MOVI    AX,0            ; Point to start
        MOV     DS,AX           ; of memory.
        MOV     AX,CS           ; Get code segment.
; Type 32
        MOVI    80H,DEV1        ; Offset for DEV1 ISR
        MOV     82H,AX          ; and segment address
; Type 33
        MOVI    84H,DEV2        ; Offset for DEV2 ISR
        MOV     86H,AX          ; and segment address
; Type 255
        MOVI    3FCH,DEV3       ; Offset for DEV3 ISR
        MOV     3FEH,AX         ; and segment address
```

Using macros

Macros should not be confused with subroutines. The essential difference between the two is that the macro is used to define code which is inserted during assembly into the program wherever the macro is named. Subroutines, on the other hand, only appear in the code once and execution is diverted to them by means of the CALL and RETURN instructions. The difference between macros and subroutines is illustrated in Figures 51 and 52.

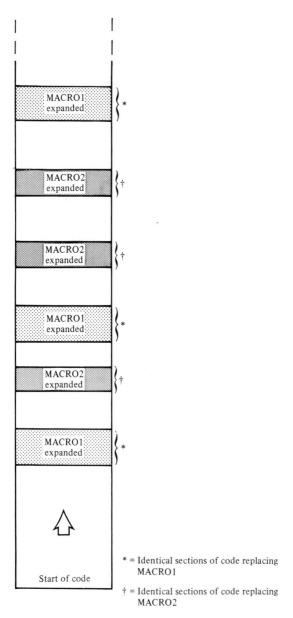

Figure 51 *Program incorporating macros*

Macros offer the programmer an exceptionally powerful and flexible tool which can be used to develop customized process control applications. In effect, the programmer is able to develop a dedicated language

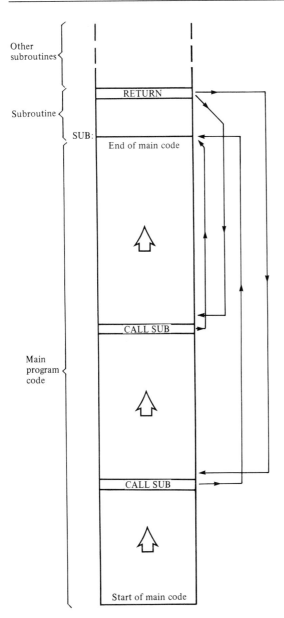

Figure 52 *Program incorporating subroutines*

defined in terms of a number of macros. Thereafter, programs can largely be written in terms of macros. Indeed, once a set of macros have been developed for a particular application, the task of the programming will have been simplified to the extent that less experienced personnel will be

138 Assembly language programming

readily capable of producing, modifying or extending software.

The following sample program should give some insight as to how valuable this technique is:

```
;           CONVEYOR CONTROL PROGRAM
;
;           VERSION 1.5        22/01/88
;
;
            MACLIB     MOTDRIVE      ; Use motor drive library
;
            SYSINIT                  ; Initialise
CYCLE:      INDICATE   READY
            WAITKEY    GO
;                                      Move load on conveyor
            DELAY      4
            INDICATE   START
            CWSTEP     20,SLOW
            CWSTEP     160,FAST
            CWSTEP     40,SLOW
            DELAY      2
            INDICATE   STOP
            WAITKEY    GO
;                                      Return empty conveyor
            DELAY      4
            INDICATE   START
            ACWSTEP    220
            DELAY      2
            INDICATE   STOP
;                                      Start over again
            RESUME     CYCLE
```

Written exclusively in terms of macros, this program is almost self-explanatory without any comments. The macro library, MOTDRIVE. LIB, contains definitions for each of the macros used together with others which may be required by other programs. The macro definition for WAITKEY, for example, may take the following form:

```
;
;           MOTOR DRIVE MACRO LIBRARY
;
;           EQUATES
;
GO          EQU        0FDH          ; Keypad code for GO
KEYPORT     EQU        1FFH          ; Keypad input port address

;
;           MACRO TO WAIT FOR OPERATOR INPUT VIA KEYPAD
;
;           PARAMETERS PASSED:   KEY
;
;           REGISTERS AFFECTED:  A, F
;
WAITKEY     MACRO      KEY
GETKEY:     LD         BC,KEYPORT    ; Get code for
            IN         A,(C)         ; key pressed
            CP         GO            ; Is it GO?
            JR         NZ,GETKEY     ; If not, keep trying!
            ENDM
```

Macros can be particularly useful in achieving a high degree of portability for a particular control program. A program written in terms of macros can be used on a variety of systems by simply changing the macro library to suit each. Hardware dependent features, such as port addresses, being defined within each macro library.

Error checking and input validation

Error checks and input validation routines should be incorporated whenever data is input and before the system accepts the data for processing. Error handling routines should be incorporated to warn the user that a fault has occurred and indicate from which source the error has arisen. This caveat also applies to operator input; an unaccepable input should be echoed to the user together with the range of acceptable responses. Care should be exercised when inputs are defaulted. The default response should result in inactivity rather than any form of positive action on the part of the system. Furthermore, the program should demand confirmation where a response or input condition will produce an irreversible outcome.

Testing

It is only possible to claim that a program has been validated after exhaustive testing in conjunction with the target hardware system. In many cases it may be possible to test individual code modules before they are linked into the final executable program. This may be instrumental in reducing debugging time at a later stage.

Testing the completed program requires simulating all conditions that can possibly arise and measuring the outcome in terms of the program's response. A common error is that of only presenting the system with a normal range of inputs. Comprehensive testing should also involve the simulation of each of the following:

1. Unexpected or nonsensical responses from the operator or user.
2. Failure of hardware components (including transducers, signal conditioning boards, disk drives, etc.).
3. Out-of-tolerance supplies (including complete power failure).

Documentation

Programmers are usually woefully lacking where program documentation is concerned. Documentation, which is essential to making the program understandable, takes various forms, the most obvious of which is the comment included in the lines of source code text.

Comments

Comments should explain the action of the source code within the program as a whole and, since the function of the operation code and operand will usually be obvious (or can be found by referring to the instruction set) there is no point in expanding on it. Comments should be reasonably brief (to save space in the source code file) but not so brief that they become cryptic. Also, there is no need to attempt to confine a comment to a single statement line. Comments can be quite effective if they read clearly and are continued over several statements to which they refer.

Headers

Headers are extended comments which are included at the start of a program module, macro definition, or subroutine. Headers should include all relevant information concerning the section of code in question and should follow a standard format. As a minimum, the following should be included:

1   Name and purpose of subroutine or macro.
2   Brief explanation of the action of the code (in terms of parameters passed, registers involved, etc.).
3   Names of other subroutines or macros on which the module depends and, where applicable, names of relevant macro libraries in which definitions are held.
4   Entry requirements (in terms of register and/or buffer contents before the module is executed).
5   Exit conditions (in terms of register contents, buffers and flags after the module has been executed).
6   List of registers used during execution of the code (which may have their contents changed as a consequence).

When producing a program header, it is wise to include any information which may be required by another programmer who may subsequently need to debug or modify the code. Nothing should be taken for granted and all loose ends should be explained.

Names

Names used for symbols and labels should be meaningful and any abbreviations used should be as obvious as possible. In the case of the names used for constants, where standard abbreviations are in common use (for example, CR for carriage return), they should be adopted. In a large program, there may be a large number of labels and/or constants and it will be necessary to distinguish between them.

As an example of the use of names, comments, and headers consider the following examples which, while functionally identical, illustrate the extremes of programming style:

```
.Z80
;               CONVERTER
;
CON:    CP          61H             ; Compare A with 61H.
        JR          C,EXIT          ; Return if carry set.
        CP          7BH             ; Compare A with 7BH.
        JR          NC,EXIT         ; Return if carry reset.
        SUB         20H             ; Subtract 20H from A.
EXIT:   RET                         ; Return.
```

Case converter subroutine

```
.Z80
;               LOWER TO UPPER CASE CHARACTER CONVERSION
;
;               ENTRY: A = ASCII character (upper or lower case)
;
;               EXIT:  A = ASCII character (upper case only)
;
;               REGISTERS: A, F
;
UPCASE: CP          'a'             ; Is it already upper case?
        RET         C               ; If so, do nothing.
        CP          'z'+1           ; Or is it punctuation?
        RET         NC              ; If so, do nothing.
        SUB         'a'-'A'         ; Otherwise, change case.
        RET
```

Improved case converter subroutine (self-documenting)

The second example shows how a program module can be made largely self-documenting by the inclusion of effective comments and a meaningful header. Note that the name of the routine has been changed so that it is easier to remember and is less likely to be confused with others. Finally, the code itself has been modified so that its action is much easier to understand.

Documentation is particularly important where software development is being carried out by several members of a team. Each development phase will rely on the documentation prepared in earlier stages, hence documentation should be considered an ongoing task and a folder should be prepared to contain the following items:

1. A detailed program specification (including any notes relevant to the particular hardware configuration required).
2. Flowcharts or descriptions of the program written in structured English.
3. Lists of all definitions and variable names.
4. Details of macro or subroutine libraries used.
5. Details of memory usage.
6. A fully commented listing of the program (latest version).
7. A diary giving the dates at which noteworthy modifications are made together with details of the changes incorporated and the name of the programmer responsible.
8. A test specification for the program with descriptions and results of diagnostic checks performed.

6
BASIC programming

Despite its limitations, BASIC remains an extremely popular language in the field of industrial process control. The continuing popularity of the language (in the face of increasing competition from arguably superior languages) is primarily attributable to the fact that the language is extremely easy to learn and that programs can be very quickly developed by those with little previous programming experience.

Programs are not only easy to produce in BASIC but they are also very easy to debug. The BASIC environment (whether interpreted or compiled) readily permits interactive debugging. Furthermore, a number of excellent implementations of the language have been developed with control applications in mind. These versions of the language also encourage programmers to produce structured code by allowing them to develop their own user-defined functions and procedures.

BASIC often exists as an interpreted language but compiled versions of the language have also become available. These produce machine code which runs many times faster than the equivalent interpreted code and offer an excellent compromise to the programmer who requires the speed and power of assembly language coupled with the ease of use offered by conventional interpreted BASIC.

Since the majority of readers will have at least a passing acquaintance with the BASIC programming language, we shall deal only with topics which are directly relevant to the development of efficient programs for control applications. Readers with no previous knowledge are advised to consult one of the many tutorial books aimed at newcomers to BASIC programming. There is no shortage of material to choose from and most texts will provide a more than adequate introduction to the subject.

The essential features to consider when purchasing a BASIC interpreter or compiler for control applications are:

1   What constructs are provided to facilitate the writing of structured code?
    Such constructs may take various forms but should typically include the ability to handle user-defined functions and procedures (with or without local variables) and such BASIC structures such as REPEAT. . .UNTIL and WHILE. . .WEND.
2   What provision is there for handling I/O?
    Most BASICs provide functions and statements (i.e., PEEK and POKE) which facilitate direct access to memory. A BASIC for control applications should also have equivalent statements that allow reading from and writing to I/O port addresses. Functions such as INP (port), and statements such as OUT port,data make writing I/O routines extremely easy.
3   Is the language suitable for generating stand alone applications (i.e., can code be stored in ROM)?
    A facility for transferring programs to ROM should be available and code developed in this way should run independently of the disk-based environment used for program development. In the case of an interpreted BASIC, a ROM-based version of the BASIC interpreter will be required. Where the language is compiled rather than interpreted, a means of replicating the run-time package in ROM must be provided.
4   Will the language allow multitasking?
    In control applications, the ability to support multitasking is a highly desirable feature. In addition to the main program, the programmer will then be able to define one or more background tasks to run concurrently with the main program. These tasks will be switched to repeatedly during program execution and thus effectively run in parallel with the main program.
5   How easy is it to interface machine code modules?
    A facility for interfacing machine code modules may also prove to be invaluable in a number of practical control applications. These modules may be crucial to the correct operation of time critical processes or may be instrumental in minimizing memory usage where program storage space is restricted.
6   What provision, if any, is there for handling interrupts?
    Some mechanism for allowing users to incorporate their own interrupt handling routines will be essential in most control applications.
7   Finally, what provision is there for error trapping?
    The ability to include specific error trapping routines can be important in making the program robust and suitable for non-technical users. Error handling routines should permit meaningful error reporting as well as the ability to retain control of the program

with an orderly shutdown when operation cannot continue.

If you can answer 'yes' to each of the foregoing questions, you can be assured that the version of BASIC in question is well suited to the control environment. Coupled with a straightforward text editor and disk/ROM filing facilities, the language should cope with almost anything.

As an example of a modern BASIC designed with control applications in mind, Appendix 9 summarizes the commands, compiler directives, functions and statements found in the popular multitasking BASIC compiler known as BASIC-Z80 (B80). This compiler contains a number of features which make it ideal for use in the STE bus environment including the ability to produce stand alone programs for storage in ROM.

Developing BASIC programs

Since it is so easy to write and enter BASIC programs it is unfortunately all too easy to develop bad habits. Furthermore, the end result produced by an unstructured program may be indistinguishable from that produced by a program which is highly structured. The difference only becomes important when the time comes for extending or modifying the program. With structured code this is a relatively simple matter. An unstructured program, on the other hand, may be a tangled nest of haphazard code and a major modification to the program may well result in the need for a complete rewrite. This can hardly be described as efficient.

There are a number of techniques which can assist in the production of efficient structured code. First and foremost, it is vitally important to get into the habit of being consistent in the layout of your programs and in the names used for variables. Failure to do this will make it extremely difficult to port sections of code from one program to another. This is a highly desirable feature which will save many hours of work. An efficient procedure for, say, accepting keyboard input and verifying that it is numeric, truncating it to integer and confirming that it is within a given range, can be useful in a huge variety of control applications. There is absolutely no reason why an efficient code module that performs such a function should not be included in every program that you write. Once written, you need never do it again.

Appendix 10 contains a selection of general purpose procedures and user-defined functions written for use with BASIC-Z80 (see Appendix 9) but can readily be modified and/or extended for use with other dialects of BASIC. Readers are free to use these routines to form the basis of their own procedure libraries.

The adoption of a consistent standard for line numbers and variable names can also be instrumental in simplifying the process of developing BASIC programs. This will permit the exchange of standard modules between programs without the need to renumber or make extensive changes to the variable names employed. We shall consider each of these topics separately.

Line numbering

A common mistake of novice BASIC programmers is that of allowing line numbers to govern the flow of the program. This accords them a significance which is not really deserved. A symptom of over reliance on line numbers is the over use of the GO TO statement, with frequent jumps backwards and forwards in the code.

At the other extreme, a common mistake is that of using a renumbering utility to renumber the entire program in strict increments (usually 10). Any significance that the line numbers may have had will be totally destroyed by such an action.

Since most versions of BASIC require line numbers, they should be used to aid program readability and simplify the process of exchanging modules between programs. All that is required is the adoption of a standard numbering scheme so that the programs follow a similar layout with the same range of line numbers used for comparable sections of code.

If this concept is new to you, it may be useful to think of each program as being analogous to a book comprising a number of chapters. Each chapter will start at a line number which is an integer multiple of 1000. Within each chapter will be found a number of subtopics, each starting at a line number which is an integer multiple of 100. Within each subtopic there will be a number of paragraphs. Each paragraph will begin with a line number which is an integer multiple of 10.

A book usually begins with some form of introduction which sets the scene and makes sure that the reader feels 'at home' with the subject. In programming terms this is referred to as initialization. Subsequent chapters then deal with specific subjects, each topic having been briefly mentioned in the introduction. Chapter 1 is particularly important as it whets the reader's appetite and leads the reader into the more specialized material that follows. Within a computer program, there will be a corresponding primary control loop or main menu. This will contain options which branch to more specialized parts of the program.

The line numbering scheme given in Table 27 is that adopted by the author and should cope with most eventualities:

Table 27

Line number range	Function
1–9	Program information (including name, copyright information, version number, date of last modification) contained in REM statements.
10–99	Program initialization (including dimensioning of arrays, calls to user-defined procedures which initialize I/O ports, and constant definitions).
100–199	Main menu display and option selection or primary program control loop.
200–899	Subsidiary menu displays or control loops (each starting on line numbers which are an integer multiple of 100).
900–999	Exit and close-down routines.
1000–1999	Subroutines and procedures unique to the application concerned.
9000–9999	Error handling routines.
10000–18999	Standard subroutines and procedures (extracted from a library – see Appendix 10).
19000–19999	Data statements.

Variable types

Wherever possible, integer variables should be used in order to minimize storage space and increase processing speed. Floating point variables, which have considerable processing and storage overhead, should be avoided. Integer variables are normally recognized by a trailing %. Thus t represents a floating point numeric variable while t% represents an integer numeric variable and t$ represents a string variable.

Integer variables require 2 bytes for storage and values can be whole numbers (i.e., no decimal points) ranging from -32768 to $+32767$. Floating point variables (floats) usually require 4 bytes for storage and values range from $-1e38$ to $1e+38$.

String variables comprise a sequence of characters (letters, numbers and punctuation) and the maximum length allowed is generally determined by the size of memory space reserved for string storage.

Variable names

In order to aid readability it has become fashionable to use relatively long names for variables. Long variable names do, however, consume valuable memory space particularly when they may appear many times in an interpreted BASIC program. In control applications, memory space is often at a premium and thus programmers can be justified in using relatively short variable names at the expense of readability.

Where a compiled BASIC is used, the overhead associated with long variable names applies only to the source files. It is thus permissible to use more meaningful variable names in such applications. Whether or not one is using a compiled BASIC, it is essential to maintain consistency with the choice of variable names. Where short variable names have to be employed, the following scheme is recommended as a starting point:

Transient numeric variables

a, b	General use.
a%, b%	General use.
c%, d%	Loop control variable (count).
e%	Last error code.
f%	Control flags.
h%	File specifier (general).
i%	File specifier (input).
k%	Key code.
l%	Console line number.
n%	General use (count).
o%	File specifier (output).
p%	Current print position.
s, t	General use.
s%, t%	General use.
u%	Console column number.
v–z	General use.
v%–w%	General use.

Numeric constants

da%	Day code.
dly%	Keyboard delay.
col%	Console columns.
ho%	Hour code.
lin%	Console lines.
lim%	Limit value.
max%	Maximum value.
mi%	Minute code.
min%	Minimum value.
mo%	Month code.
se%	Seconds code.
sx%	Screen width (characters).
sy%	Screen height (characters).
ver	Version number.
we%	Week code.
ye%	Year code.

Transient string variables

a$–d$	General use.
f$	Filename.
i$	User input (validated).
n$	User name or identifier.
p$	Prompt string.
r$	User response (keyboard input).
s$–z$	General use.

String constants

al$	Alert message.
cl$	Console message.
dat$	Date (mm/dd/yy).
day$	Day of the week.
err$	Error message.
tim$	Time (hh/mm).
mon$	Month.
nam$	Company name.
ver$	Version message.
war$	Warning message.
yer$	Year.

Subroutines

Subroutines can be instrumental in making very significant reductions in the size of BASIC programs and they should be employed whenever a section of code is to be executed more than once. Note, however, that if your version of BASIC supports the use of procedures or user-defined functions then these should be used instead. A typical example of the use of a subroutine might involve a delay routine which is required at various points in a program. Assuming that such a routine was located at line 10100, it might take the following form:

```
10100 REM Delay subroutine
10110 FOR c%=0 TO 10000: NEXT c%
10120 RETURN
```

The subroutine may be called from several points within the main program as follows:

```
340 ..........
350 GOSUB 10100
360 ..........
```

etc.

```
440 ..........
450 GOSUB 10100
460 ..........
```

etc.

```
710 ..........
720 GOSUB 10100
730 ..........
```

In each case, program execution resumes at the line immediately following the GOSUB statement. Also note that, on exit from the subroutine, c% will have the value 10000.

We could make the delay subroutine even more flexible (allowing for variable length delays) by altering the upper limit of the loop using a variable which is set immediately prior to the subroutine call. The modified subroutine would then become:

```
10100 REM Delay subroutine
10110 FOR c%=0 TO lim%: NEXT c%
10120 RETURN
```

As before, the routine may be called from several points in the main program as follows:

```
340 ..........
350 lim%=10000: GOSUB 10100
360 ..........
```

etc.

```
440 ..........
450 lim%=20000: GOSUB 10100
460 ..........
```

etc.

```
710 ..........
720 lim%=15000: GOSUB 10100
730 ..........
```

On exit from the subroutine, the value of c% will have been modified to whatever lim% was set to immediately prior to the subroutine call.

Procedures

A user-defined procedure can be thought of as a named subroutine. The procedure is simply called by name rather than by GOSUB followed by a line number. This can be instrumental in not only making the resulting code more readable but it also ensures that the structure of the program can be easily understood. A further advantage of procedures is that parameters may be passed into procedures and values returned to the main program. Variables used only within the context of the procedure (i.e., not returned to the main program) should be declared as LOCAL.

Procedures are defined using statement of the form DEF name or DEF PROC name and are terminated by statement of the form ENDPROC. ENDPROC name, or RETURN. Procedures may also contain references to other procedures (i.e., procedures can be 'nested'). Procedure names should be chosen so they do not conflict with any variable names nor should they be BASIC reserved words.

The previous delay subroutine can be easily written as a procedure:

```
10100 REM Delay procedure
10110 DEF delay(lim%)
10120    LOCAL c%
10130    FOR c%=0 TO lim%: NEXT c%
10140 ENDPROC
```

The method of calling the delay procedure is more elegant than that used with the equivalent subroutine and takes the following format:

```
340  ..........
350  delay(10000)
360  ..........
```

etc.

```
440  ..........
450  delay(20000)
460  ..........
```

etc.

```
710  ..........
720  delay(15000)
730  ..........
```

The values within parenthesis are parameters passed into the procedure as lim%. Such values are local to the procedure and external references to lim% will remain unchanged by the action of the procedure.

User-defined functions

User-defined functions are similar to user-defined procedures but return values (integer, float, or string) to the main program. As with user-defined procedures, functions are called by name (or FN name) but are usually terminated by an equivalence sign followed by the expression or variable which specifies the value to be returned to the main program.

Logical constructs

Modern BASICs provide us with a number of other useful constructs which can be instrumental in the production of efficient structured code. As an example, a somewhat more elegant delay procedure can be produced using the WHILE...WEND construct. This routine uses a single variable rather than the two that were required in the FOR... NEXT construct used earlier.

```
10100 REM Delay procedure
10110 DEF delay(lim%)
10120    WHILE lim%>0
10130       lim%=lim%-1
10140    WEND
10150 ENDPROC
```

The condition in the WHILE stated is tested and, as long as it remains true (i.e., evaluates to non-zero), the code within the loop will be repeated.

Prompts and messages

Any program to be used by a person other than the originator should incorporate meaningful prompts and messages to aid the user. Prompts should also give some indication of the input required from the user in terms of the acceptable keystrokes, the length of an input string, and the need to include a RETURN keystroke. The following are examples of acceptable prompts:

Do you wish to quit? (Y/N)
Press [SPACE] to continue . . .
Enter today's date (MM:DD:YY) followed by [RETURN] . . .
Enter filename (maximum 8 characters) followed by [RETURN] . . .

Messages, unlike prompts, demand no immediate input from the user and should be included at any point in the program at which the user may require information concerning the state of the system. Messages should be written in plain English and should not assume any particular level of technical knowledge on the part of the user. The following are examples of acceptable messages:

Loading data file from disk . . . please wait!
Printer is not responding – please check paper.
Warning! Transducer on channel 4 is not responding.

Keyboard entry

Keyboard input from the user will be required in a variety of applications.

Such input may take one of three basic forms summarized below:

1 Single keystrokes. Keystrokes may either be a letter, number, or punctuation and will generally not require the use of the RETURN or ENTER key.
2 Numerical inputs (comprising one or more keystrokes terminated by RETURN). Each keystroke must be a number (or decimal point in the case of floats) and the input will normally be assigned to a numeric variable (either integer or floating point).
3 String inputs (comprising one or more keystrokes terminated by RETURN). Each keystroke may be a number, letter or punctuation. The string input by the user will normally be assigned to a string variable.

Single key inputs

Single key inputs will be required in a wide variety of applications. Such inputs can take various forms including menu selections or simple 'yes/no' confirmations. In either case, it is important to make the user aware of which keys are valid in each selection and, where the consequences of a user's input is irrevocable, a warning should be issued and further confirmation should be sought.

A simple typical 'yes/no' dialogue would take the following form:

```
550 INPUT "Are you sure (Y/N) "; r$
560 IF r$="Y" THEN ... ELSE ...
```

This piece of code has a number of shortcomings not the least of which is that it will accept any input from the user including a default (i.e., RETURN or ENTER used on its own). Other problems are listed below:

1 The user may not realize that the input has to be terminated by ENTER or RETURN.
2 A response of 'N' is not distinguished from a default (or any input other than 'Y').
3 The routine does not allow a lower case input and the user may not realize that the SHIFT key has to be applied.
4 If the user replies with 'YES' or 'yes', this would be equivalent to 'N'!
5 Finally, since we would probably want to use the routine at several points within the program, it should be coded as a procedure or user-defined function.

A much better solution to the problem would take the following form:

```
550 confirm
560 IF f% THEN ... ELSE ...
```

The user-defined function (confirm) would then be written along the following lines:

```
10900 REM Get confirmation flag from user
10910 DEF confirm
10920    LOCAL k%
10930    f%=-1
10940    PRINT "Are you sure ? (Y/N)"
10950    REPEAT
10950       k%=KBD
10960       IF k%=&59 OR k%=&79 THEN f%=1
10970       IF k%=&4E OR k%=&6E THEN f%=0
10980    UNTIL f%+1
10990 =f%
```

The function returns a flag, f%, which is true (non-zero) if the user presses 'Y' or 'y' and is false (zero) if the user presses 'N' or 'n'. The KBD function waits until a character is available from the keyboard and then returns its ASCII value in k%. Note that KBD, unlike INPUT, does not require the use of the RETURN or ENTER key as a terminator.

The REPEAT...UNTIL loop ensures that k% has a value which corresponds to one of four acceptable values (corresponding to keys 'Y', 'y', 'N' and 'n'). Any other keyboard input is invalid and the program continues to wait for further keyboard input until an acceptable value is returned (during this time the prompt message remains on the screen and does not scroll).

Now, to consider a more complex example, let us consider the case of a main menu selection. Suppose we are dealing with a control system which has four main functions (each of which is to be handled by a secondary menu) together with a function which closes down the system and exits from the program.

To conform with our established line number convention, the main menu should start at line 100 and subsidiary menus should start at lines 200, 300, and so on.

The five main functions will be as follows:

1 Set parameters.
2 Heater control.
3 Pump control.
4 Print report.
5 Close down.

The following code can be used for the main program loop:

```
100 REM Main menu selection
105 WHILE 1
110    clear_screen
120    PRINT "        MAIN MENU"
121    PRINT
122    PRINT "  [1] Set parameters"
```

```
123     PRINT "  [2] Heater control"
124     PRINT "  [3] Pump control"
125     PRINT "  [4] Print report"
126     PRINT "  [5] Close down"
127     PRINT
128     PRINT " Option required (1 - 5)?"
129     scroll(6)
130     REPEAT
131        k%=KBD
132        UNTIL k%<&36 AND k%>&30
132     IF k%=&31 THEN set_param
133     IF k%=&32 THEN heater_control
134     IF k%=&33 THEN pump_control
135     IF k%=&34 THEN report
136     IF k%=&35 THEN close_down
195     WEND
```

Notice that the main program loop consists of an infinite WHILE. . . WEND loop. The integer value returned by the KBD function is then tested within a REPEAT. . .UNTIL to see whether it is within range of the valid keyboard responses (note that depressing the '1' key returns an ASCII value of 31H, and so on). The REPEAT. . .UNTIL loop is only exited when a valid keystroke is detected. Having obtained a valid keystroke, the program checks the response to see which key was depressed using a series of IF. . .THEN statements so that the desired procedure can be called.

If, for example, the user had pressed the '5' key, the result of the IF. . .THEN statement in line 136 would have been found to be true (all others having been false) and program execution would be diverted to the user-defined procedure named close_down. In this case, and since the result of the close_down routine is irrevocable, the user should be given the option of returning to the main menu. Hence the close_down procedure should take the following form:

```
900  REM Close down and exit
910  DEF close_down
920     clear_screen
930     PRINT "You have selected the CLOSE DOWN option."
940     scroll(4)
950     confirm
940     IF f%=0 THEN ENDPROC
950     ..........
```

etc.

As before, the confirmation function returns a flag, f%, which is true (non-zero) if the user presses 'Y' or 'y' but is false (zero) if the user presses 'N' or 'n'. If the user decides not to continue with the close_down procedure, RETURN ensures that the procedure is abandoned and execution resumes at line 195 (the line following the procedure call). The WEND statement then diverts the program back to the beginning of the main menu selection routine.

Numerical inputs

The simple method of dealing with numerical input involves using a BASIC statement of the form:

```
650 INPUT "Value required"; n%
```

This line of code will only work properly if the user realizes that a numeric value is required. Since BASIC cannot assign a letter to a numeric variable, the program will either crash or assign a value of zero if the user inadvertently presses a letter rather than a number. Furthermore, it would be useful to be able to impose a range of acceptable values on the user. The program should reject input values outside this range, warn users that their input is invalid, and prompt again for further input. Again, such a routine would be ideally coded as a procedure. The procedure call could typically take the form:

```
650 prompt$="Temperature required"
660 number_in(60,90)
```

while the procedure itself would be coded along the following lines:

```
11000 REM General purpose integer numerical input
11010 DEF number_in(min%,max%)
11020    PRINT prompt$;" ";
11030    INPUT n$
11040    n%=VAL(n$)
11050    IF n%<=max% AND n%>=min% THEN ENDPROC
11060    PRINT "Value outside permissible range!"
11070    GOTO 11020
```

The procedure prints the prompt string (prompt$) and assigns the user's input to a string variable in order to avoid the program crashing if a letter is inadvertently pressed. The string is subsequently converted to an equivalent numeric variable using the VAL function. The resulting integer is then tested to see whether it lies within the acceptable range. If the integer is within range, the procedure is exited with n% containing a valid integer input. If the integer is not within range, the user is warned and prompted for further input. A similar routine can be produced for floating point input and, if desired, the prompt string can be included in the list of parameters to be passed into the function.

String inputs

The simple method of dealing with string input involves using a BASIC statement of the form:

```
650 INPUT "Name"; n$
```

This line of code is fortunately not quite so prone to problems as its equivalent for numeric input. It is, however, worth considering what action we should take if the user should default the input (i.e., just presses RETURN or ENTER) or proceeds to input an unacceptably long string (the latter is an important consideration when dealing with filenames). Hence our general purpose string input routine should allow for the substitution of a default string and should also truncate the user's input to a specified length. The procedure call might take the following form:

```
650 prompt$="Filename"
660 string_in(8,"CONTROL")
```

While the procedure itself would be coded along the following lines:

```
11100 REM General purpose string input
11110 DEF string_in(length%,default$)
11120    PRINT prompt$;" ? ";
11130    INPUT LINE r$
11140    IF r$="" THEN r$=default$
11150    shorten(r$,length%)
11160 ENDPROC
```

As before, the procedure prints the prompt string (prompt$) and assigns the user's input to a string variable. The use of INPUT LINE (rather than just INPUT) ensures that the user can include punctuation. The user's response (r$) is then checked to determine whether it is a null string (i.e., the user has defaulted) and, if so, the specified default string is substituted. Finally, a further procedure (shorten) is called to truncate the string to the specified length. This procedure takes the form:

```
10700 REM Shorten string to a specified length
10710 DEF shorten(long$,length%)
10720    string$=LEFT$(long$,length%)
10730 ENDPROC
```

Table 28 gives typical user entries and resulting values returned to the main program (in string$) by the foregoing code:

Table 28

User input	Value returned
CON1	CON1
CONTROLDAT	CONTROLD
(default)	CONTROL

Data files

Finally, the ability to store data acquired by a control system is important where a detailed analysis of system performance is required. Data may be stored in one or more disk files in a disk-based system. Such files can readily be manipulated from BASIC.

The stages required for saving data in a disk file are as follows:

1 Open the file for output (using OPENOUT) and include a filename or complete file specification. The operating system will return an associated channel number which should be used for subsequent operations on the file.
2 Send data to the file (using PUT# or PRINT#).
3 Close the file (using CLOSE#).

As an example, let us assume that we have an integer array of thirty-two floating point values, a( ), to be stored in a disk file. If the data file is to be called 'temp.dat' and is to be stored on the default drive, the following code could be used:

```
7000 REM Save current data in disk file
7010 DEF save_dat
7020   clear_screen
7030   PRINT "Saving data on disk - please wait!"
7035   scroll(4)
7040   o%=OPENOUT("temp.dat")
7050   FOR n%=0 TO 31
7060     PRINT#o%,a(n%)
7070   NEXT n%
7080   CLOSE#o%
7090 RETURN
```

The stages required for loading data from a disk file are as follows:

1 Open the file for input (using OPENIN) and include a filename or complete file specification. The operating system will again return an associated channel number which should be used for subsequent operations on the file.
2 Retrieve data from the file (using GET# or INPUT#).
3 Close the file (using CLOSE#).

The following code can be used to retrieve the data stored by the previous example, loading it back into array a( ):

```
7100 REM Load data from disk file
7110 DEF load_dat
7120   clear_screen
7130   PRINT "Loading data from disk - please wait!"
7135   scroll(4)
7140   i%=OPENIN("temp.dat")
7150   FOR n%=0 TO 31
7160     INPUT#i%,a(n%)
7170   NEXT n%
7180   CLOSE#i%
7190 ENDPROC
```

7
C programming

C was first developed as a tool for writing system software. As such it is more than just a language for generating applications; it provides a standard environment for software development. In recent years, C has increased in popularity among systems and applications programmers – many of whom readily admit to being both captivated and confounded by the language.

Whereas C has obvious advantages in the system world (it lends itself well to structured programming, is eminently portable, has a relatively small 'core', is compact and offers execution speeds which are far in excess of those which can be obtained with interpreted languages) its advantages in control applications may be less obvious.

To appreciate the advantages of C in a control environment, consider the process of producing an executable program from source text written in C. The source file (ASCII text written in C) is first compiled into assembly language source code. Thereafter, the resulting code is assembled into object code. It is thus eminently possible to include assembly language code within the text of a C program. The ability to include in-line assembly language code is arguably the single most important consideration in favour of using C in the context of process control applications. The programmer can thus take advantage of C as a rapid means of producing programs written in a (relatively) high-level language while at the same time having the freedom to incorporate the lower level assembly language code necessary for interaction with hardware. Furthermore, since C provides a reasonably standard environment, it is eminently possible to transfer programs from one hardware configuration to another. One only needs to make changes to the hardware specific I/O routines.

To the newcomer, program source text written in C can appear somewhat cryptic. Furthermore, programmers experienced in other, less structured languages may have difficulty when making the transition to

C. Indeed, it is often said that it is easier to learn C if one has not had the misfortune of acquiring preconceptions developed as a result of familiarity with BASIC. While this may be demonstrably true, the fact is that most of today's learners of C will already be proficient in one or more other languages and these will invariably include BASIC.

C/80

The examples given in this chapter are written in C/80 produced by Software Toolworks. This compiler, which runs under the CP/M operating system, has a number of features which make it ideally suited to developing software for control applications. The compiler requires a minimum of 56 K RAM and produces assembly language text files which are assembled into executable object code by the AS absolute assembler. C/80 can also produce object code output for Microsoft's M80 or Digital Research's RMAC relocatable assembler.

C/80 version 3.1 supports all of the standard features of C defined in the C programming language (see Bibliography) with the following exceptions:

- Float and long data types (these can be added using C/80 MATHPAK).
- Double data type.
- Typedef.
- Arguments to define macros.
- Bit fields.
- Line.
- Declarations within nested blocks.

C/80 also provides the following features:

- UNIX-style I/O redirection and command line expansion.
- Conventional C I/O and string library.
- Formatted and random access file I/O.
- Dynamic storage allocation.
- Selectable M80 or RMAC compatability.
- In-line assembly language.

Invoking the C/80 compiler

The C/80 compiler is invoked by commands of the form:

C filename

which generates an assembly language file having the same name as the C source file but with an ASM extension. As an example, the command:

C PROCESS

will take a C source file named PROCESS.C and produce an assembly language file named PROCESS.ASM.

If desired, the command can use a complete file specification rather than a filename. The command:

C B:PROCESS.MAC = A:PROCESS.C

will take a C source file named PROCESS.C from drive A and produce an assembly language file named PROCESS.MAC on drive B.

Invoking the AS assembler

Having compiled a C source file with no errors, the AS assembler is used to assemble the program into an executable program. During this process, the C run-time library (CLIBRARY.ASM) should be resident on the disk placed in drive A together with the AS.COM program file.

The AS absolute assembler is invoked by commands of the form:

AS filename

which generates an executable program file having the same name as the assembly language file but with a COM extension. As an example, the command:

AS PROCESS

will take an assembly language file name PROCESS.ASM and produce an executable program file named PROCESS.COM.

The command may also use a complete file specification rather than a filename. The command:

C B:PROG1.COM = A:PROCESS.ASM

will take an assembly language source file named PROCESS.ASM from drive A and produce an assembly language file named PROG1.COM on drive B.

C programming techniques

Rather than devote space in this chapter to introducing readers to the basic concepts associated with C programming, we shall adopt the same approach to that used in Chapter 6 by providing a tutorial aimed specifically at showing how C can be used in control applications. Topics have therefore been included which have particular relevance to process control. Newcomers to the language are advised to refer to one or more of the texts listed in the Bibliography prior to, or concurrently with, reading this chapter.

Functions

As readers will doubtless be aware, the fundamental building blocks of C programs are called functions. Once written, functions (like BASIC procedures) may be incorporated in a variety of programs whenever the need arises. The following function definition provides a delay:

```
delay()
{
   int x;
   for (x=1;x<=20000;++x);
}
```

The delay function is called from a main program by a statement of the form:

```
delay();
```

It is important to note that no semi-colon follows the closing bracket of the function definition whereas, when the function is called, the program statement is terminated, as usual, by a semi-colon. The main body of the function is enclosed between curly braces ({ and }). Since C is essentially a 'free-form' language (i.e., the compiler ignores white space within the source text) programmers are able to adopt their own style of layout within the source text. The C functions and programs presented in this chapter will, however, follow the author's recommended layout convention summarized below:

1. Matching opening and closing braces ({ and }) are vertically aligned with one another.
2. Statements within the body of a function are indented by three columns with respect to their opening and closing braces.
3. Expressions (enclosed in brackets) used in conjunction with for and while statements are placed on the same line as the matching for or while.
4. Blank lines are used to separate function definitions.
5. The first function defined in a program is main( ).

Returning to the previous example, readers will probably have spotted a fundamental weakness in the simple delay function arising from the fact that it is only capable of providing a fixed delay. The function can be made more versatile by passing a parameter into it. The following modified delay function achieves this aim:

```
delay(limit)
int limit;
{
   int x;
   for (x=1;x<=limit;++x);
}
```

The argument (contained in parenthesis after the function name) is defined as an integer type before the function body. The function is then called using a statement of the form:

```
delay(20000);
```

Thereafter, the value 20000 is passed to the function and is used as the value for limit. Where more than one argument is required, they are simply listed and separated by commas. The data type for each argument must then be defined before the opening brace of the function body. A function definition dealing with port output, for example, might be declared with statements of the form:

```
out(port,byte)
int port, byte;
{
......
......
}
```

The corresponding function call would require a statement of the form:

```
out(255,128);
```

In this case, the value 255 would be passed into port while the value 128 would be passed into byte.

Messages

Messages in C can be printed using statements of the form:

```
printf(message_string);
```

The standard C printf statement is, however, more versatile than its equivalent in BASIC and it allows a wide variety of formatting variations. Those available in the C/80 implementation include:

\b for backspace
\f for form feed
\n for new line
\t for tab

The following example prints the message 'Warning!' immediately preceded and immediately followed by two blank lines:

```
printf("\n\nWarning!\n\n");
```

Variables can be included within the formatted print statement, as the following example shows:

```
printf("Tank number %d temperature %d\n",tankno,temp);
```

164 C programming

The current values of tankno and temp are printed as integer decimal numbers within the string. Thus if tankno and temp currently had the values 4 and 56, the resulting output generated would be:

Tank number 4 temperature 56

C allows a wide range of conversion characters to be included within formatted print strings. These usually include:

%c Single character
%d Signed decimal.
%o Unsigned octal.
%s String.
%u Unsigned decimal.
%x Hexadecimal.

The following example shows how conversion specifiers can be used to print the decimal, hexadecimal and octal value of the same number:

```
printf("Decimal %d, hexadecimal %x, octal %o",num,num,num);
```

It is important to note that each conversion specifier must correspond to an argument within the list.

Loops

Loops can be easily implemented in C programs. The following program prints the ASCII character set and uses the %d, %x, and %c conversion specifiers to provide the decimal, hexadecimal and ASCII representation of the loop index (byte):

```
#include printf.c
main()
{
  int byte;
  for (byte=32;byte<128;++byte)
  {
    printf("Decimal %d, hex., %x, ASCII %c\n",byte,byte,byte);
  }
  exit();
}
```

The loop is executed for byte values in the range 32 to 127 and byte is increased by 1 on at the end of each pass round the loop.

Loops can also be nested to any required depth. The following program provides a simple example based on the use of while rather than for:

```
#include printf.c
main()
{
   int s;
   s=0;
   while (s<4)
   {
      ++s;
      printf("Outer loop count = %d\n",s);
      inner();
   }
   exit();
}
inner()
{
   int t;
   t=0;
   while (t<4)
   {
   ++t;
   printf("   Inner loop count = %d\n",t);
   }
}
```

The outer loop is executed four times (with s taking the values 0 to 3 in the expression following while). The inner loop is executed four times (with t taking the values 0 to 3 in the expression following while) on each pass through the outer loop.

Inputs and prompts

A single character can be returned from the standard input (usually the terminal) by means of the getchar( ) function. The following routine shows how a single character can be returned from the keyboard:

```
#include printf.c
main()
{
   char c;
   c=inchar("Enter option required ...");
   printf("\n\nOption selected = %c\n",c);
   exit();
}
int inchar(prompt)
char *prompt;
{
   printf("\n%s",prompt);
   return(getchar());
}
```

Here we have defined a function, inchar, which returned an integer to main. This is automatically converted to a character and assigned to c.

166 C programming

Where a multiple (rather than single) character string is required, the scanf function can be used, as shown in the code fragment below:

```
char code[16];

getcode()
{
  printf("Enter operator code ...");
  scanf("%s",code);
}
```

The scanf function allows a similar set of conversion characters to that available for use within printf. It is important to note that scanf terminates input when a return, space or tab character is detected. Furthermore, the array must be sufficiently large to accommodate the longest string likely to be input. Since the string is automatically terminated by a null character, the array must be dimensioned so that its number of elements is one greater than the maximum string length.

Multiple arguments may be included within scanf. It is important to note that, unlike printf, the arguments to scanf are pointers (not variables themselves). This point regularly causes confusion. Finally, since scanf involves considerable overhead, simpler functions may be preferred where the space for code is strictly limited.

Menu selection

The following example illustrates the combined use of getchar, printf, and scanf in a simple decimal to hexadecimal conversion utility:

```
#include printf.c
#include scanf.c
char number[16];

main()
{
  int num, c;
  c='Y';
  printf("\n\nDECIMAL TO HEXADECIMAL CONVERSION\n\n");
  while ((c=='y')||(c=='Y'))
  {
    printf("\nEnter decimal number (max. 65535) ");
    scanf("%s",number);
    num=atoi(number);
    printf("\nDecimal %u = %x hexadecimal",num,num);
    printf("\n\nEnter [Y] to continue, any other key to exit ");
    c=getchar();
  }
  exit();
}

#include stdlib.c
```

The expression following while evaluates true if getchar returns either 'y' or 'Y' (allowing for shift or capitals lock). Any other response from the user (including a default) results in the expression evaluating false and the program returns control to the operating system.

More complex menu selections can be achieved using the switch...case statement. The following examples shows how:

```
#include printf.c
#define FOREVER 1

main()
{
   char c;
   while (FOREVER)
   {
     menu();
     c=getchar();
     switch(c)
     {
       case '1':
          init();
          break;
       case '2':
          pump();
          break;
       case '3':
          mix();
          break;
       case '4':
          deliver();
          break;
       case '5':
          exit();
     }
     c=getchar();
   }
}

scroll(nolines)
int nolines;
{
   int x;
   for (x=0;x<nolines;++x)
   {
     printf("\n");
   }
}

menu()
{
   scroll(10);
   printf("     MAIN   MENU\n\n");
   printf("[1] Initialise the system\n");
   printf("[2] Pump control\n");
   printf("[3] Mixer control\n");
   printf("[4] Delivery control\n");
   printf("[5] Close down and exit");
   printf("\nEnter option required ... ");
}
```

```
init()
{
  scroll(6);
  printf("INITIALISING SYSTEM\n");
  ......
  ......
  ......
}

pump()
{
  scroll(6);
  printf("PUMP CONTROL\n");
  ......
  ......
  ......
}

mix()
{
  scroll(6);
  printf("MIXER CONTROL\n");
  ......
  ......
  ......
}

deliver()
{
  scroll(6);
  printf("DELIVERY CONTROL\n");
  ......
  ......
  ......
}
```

The foregoing program, which should largely be self-explanatory, may readily be adapted to the user's own requirements and clearly shows how C encourages structured programming style.

Passing arguments into main

A useful facility available within C running under CP/M is that of passing arguments into programs from within the command line input by the user when the program is first loaded. The main function allows two arguments, argc and argv. When main is called, argc is the number of elements in argv, and argv is an array of pointers to the strings which appear in the command line.

```
#include printf.c
#define EOF 26

main(argc,argv)
char *argv[];
int argc;
{
```

```
  int chan, c;
  chan = fopen(argv[1],"rb");
  while (c!=EOF)
  {
    c=getc(chan);
    putc(c,0);
  }
  fclose(chan);
  exit();
}
```

The program is a simple utility which sends a named file to the console device. Assuming that the program is called SPRINT, the command line entered after the operating system prompt, would take the form:

SPRINT filename

To print a file called HELLO.DOC, the command would be:

SPRINT HELLO.DOC

More than one argument can be passed into main. The following is a simple utility program for renaming files which passes two arguments into main:

```
#include printf.c
#define EOF 26

main(argc,argv)
char *argv[];
int argc;
{
  int v;
  v = rename(argv[1],argv[2]);
  if (v == -1)
    {
      printf("Rename failed!\n");
      exit();
    }
  printf("File renamed.\n");
  exit();
}

#include stdlib.c
```

Assuming that the program is called CNAME, the command line entered after the operating system prompt, would take the form:

CNAME oldname newname

To change the name of a file called HELLO.DOC to GOOD-BYE.DOC the command would be:

CNAME HELLO.DOC GOODBYE.DOC

Assembly language

Assembly language can be very easily incorporated within C/80 functions. The following example shows how the ports of an 8255 PPI device can be initialized from a C function:

```
init()
{
  printf("Initialising ports ...\n");
#asm
       MVI  A,89H
       OUT  0E3H
       MVI  A,92H
       OUT  0E7H
       MVI  A,0FFH
       OUT  0E6H
#endasm
}
```

The compiler directives, #asm and #endasm, allow in-line assembly. The program lines between the two statements are ignored by the compiler and are merely passed unchanged into the assembly language (ASM) output file.

Disk files

File handling is quite straightforward in C. Files must be opened before use using a statement of the form:

```
fopen (filename, mode);
```

The filename can be a file specification or the name of a logical device. The mode can be 'r' for read, 'w' for write, and 'u' for update. If the file cannot be opened (for example, it is not present on the disk) fopen returns 0 otherwise fopen returns the channel number to be used in conjunction with subsequent read or write operations.

After use, files must be closed using statement of the form:

```
fclose(chan);
```

Where chan is the channel number returned by a previous fopen statement.

The final example in this chapter is a utility which can be used to send the contents of a disk file to a printer. The program opens channel fp to a disk file (specified in argv from the command line) and fq to the CP/M LST: device (i.e., the printer). The disk file is then read, character by character, until the end of file (EOF) marker is reached. As each character is read from channel fp, it is output to channel fq and also echoed to the console display.

```
#include printf.c
#define EOF -1
int i, pc, fp, fq;

main(argc, argv)
int argc;
char *argv[];
{
  fp=fopen(argv[1],"r");
  if (fp==0)
     {
       printf("Can't open %s!\n",argv[1]);
       exit();
     }
  fq=fopen("LST:","w");
  if (fq==0)
     {
       printf("Printer not responding!\n");
       exit();
     }
  printf("\nPrinting %s, please wait!");
  i=1;
  pc=1;
  while ((pc=getc(fp))!=EOF)
     {
       if((i!=60)&&(pc==32))
          {
            putc(13,fq);
            putc(10,fq);
            i=1;
          }
       else
          putc(pc,fq);
          putchar(pc);
       if (pc!=13)
          ++i;
       else
          i=1;
     }
  fclose(fq);
  fclose(fp);
  exit();
}
```

8
Interfacing

This chapter aims to introduce readers to the general principles of interfacing sensors and transducers to STEbus modules. We shall describe a variety of common sensors and transducers and, for those who do not wish to make use of 'off the shelf' signal conditioning modules, details of the circuitry necessary to interface such devices to standard STE slave I/O cards will be provided. Before embarking on this task, it is perhaps worth mentioning some of the more important characteristics and limitations of typical digital and analogue I/O ports.

Characteristics of digital I/O ports

The digital I/O ports provided by STE I/O slave cards are invariably byte wide (i.e., each port comprises eight individual I/O lines). Such ports are usually implemented with the aid of one or more of the programmable parallel I/O devices described in Chapter 2 and the I/O lines they provide are usually arranged in groups of eight within the 50-way connector commonly used for STE I/O (see Appendix 4). This connector provides for up to five groups of eight I/O lines (individual groups may be associated with different programmable parallel I/O devices) and also carries the supplies given in Table 29.

Table 29

Supply	Pin number
+ 5 V	50
+12 V	48
−12 V	47
0 V	1 and 2

Where parallel I/O devices are connected directly to the STE digital I/O connector, care should be taken to ensure that no line is excessively loaded. The B group lines of a programmable parallel I/O device are usually able to source sufficient current to permit the direct connection of the base of an NPN (preferably Darlington type) transistor. To minimize loading on the remaining I/O lines it will generally be necessary to employ the services of one or more octal TTL buffers. In any event, it is important to note that, when sourcing appreciable current, the high level output voltage present on a port line may fall to below 1.5 V. This will be acceptable when driving a conventional or Darlington transistor but represents an illegal voltage level as far as TTL devices are concerned.

Fortunately, most STE digital I/O cards incorporate buffers or transceivers. The buffered output lines that they provide can often source and sink as much as 15 mA and 24 mA respectively. Typical of devices used in such applications is the 74LS245 octal non-inverting bus transceiver. Note that, where such a device is fitted to a digital I/O card, it will be necessary to configure it as either input or output as part of the general port initialization. Furthermore, if the direction of the port is changed during program execution, it is important not to forget that the direction of the transceiver must also be altered.

A typical initialization procedure for a 40-channel digital I/O card based on two 8255 PPI devices and fitted with 74LS245 buffers might involve the following steps:

1. *Initialize PPI1*. Ports A, B and C of PPI1 should be initialized as either inputs or outputs (as required) by writing the appropriate control word to PPI1's control register.
2. *Initialize PPI2*. Ports A and B of PPI2 should be initialized as either inputs or outputs (as required) while port C should be initialized as an output since its lines are used to control the buffers fitted to the other ports. This is achieved by writing an appropriate control word to PPI2's control register.
3. *Initialize buffers*. Output the appropriate bit pattern (byte) to port C of PPI2 in order to enable the buffers on the other ports and configure them for operation in the correct direction (i.e., if port A of PPI1 has been initialized for output then its corresponding buffer must be enabled and also configured for output by setting the appropriate bits in the data byte sent to port C of PPI2).

Having configured a digital I/O port, it is a relatively easy matter to send data to it or read data from it. Each port will appear as a unique address within the STEbus I/O map and data can be read from or written to the port using appropriate IN and OUT statements. Where the digital I/O lines within a port group have individual functions, appropriate bit masks can be included in the software so that only the state of the line in question is affected during execution of an OUT command.

Characteristics of analogue I/O ports

STE analogue I/O ports generally provide up to sixteen analogue input lines and several analogue output lines. Analogue I/O ports are often based on one or more of the following devices:

ZN427E	8-bit ADC
ZN425E	8-bit DAC (single supply)
ZN428E	8-bit DAC (single supply)
ZN435E	8-bit DAC (single supply)
ZN439E	8-bit ADC (single supply
ZN448	8-bit ADC (single supply)
ZN502E	10-bit ADC
573	10-bit ADC
574	12-bit ADC
667	12-bit DAC
7226	8-bit DAC (4 channel)
7528	8-bit DAC (2 channel)
7542	12-bit DAC
7545	12-bit DAC
7581	8-bit ADC (8 channel)

The I/O lines provided on an STE analogue I/O card are often grouped together in accordance with the convention shown in Appendix 5. The 50-way connector provides for sixteen analogue inputs and up to four analogue outputs. In addition to the I/O lines, the standard I/O connector also carried the supplies given in Table 30.

Table 30

Supply	Pin number
+ 5 V	50
+ 5 V	49
+12 V	48
−12 V	47
0 V	1 and 2

Electrical characteristics

Analogue inputs generally exhibit a high input resistance (50 kΩ or more) and operational amplifier buffers are usually fitted to provide voltage gain adjustment and additional buffering between the analogue input and the input of the ADC chip.

Analogue outputs are usually available at a relatively low output impedance (100 Ω or less) and are invariably buffered from the DAC by means of operational amplifier stages. Typical output voltages produced by an analogue output port utilizing an 8-bit DAC range from 0 V to 5.1 V (20 mV/bit) when configured for unipolar operation or −5.1 V to +5.1 V (40 mV/bit) when bipolar operation is selected.

The procedure for reading values returned by an analogue input port will vary depending upon the type of ADC used. A typical sequence of operations for use with a multichannel analogue input card with 8-bit resolution based on the ZN448E ADC would take the following form:

1 Select the desired input channel and start conversion.
   Send the appropriate byte to the status latch in order to select the required channel and input multiplexer. Conversion starts automatically when data is written to the status latch address.
2 Either:
   (a) Wait 10 µs (this is just greater than the 'worst-case' conversion time) using an appropriate software delay or,
   (b) Continuously poll the ADC to sense the state of the end-of-conversion (EOC) line. This signal appears as a single bit in the status byte and, when low, it indicates that conversion is complete and valid data is available from the ADC.
3 Read the data.
   Having ensured that conversion is complete, the valid data byte can be read from the appropriate ADC address.

The byte read from the port will take a value between 00H and FFH. If the ADC has been configured for unipolar operation, a value of 00H will correspond to 0 V while a value of FFH will correspond to full-scale positive input (typically 5.1 V). When bipolar operation is used, a data byte of 00H will indicate the most negative voltage (typically −5.1 V) while FFH will indicate the most positive voltage (typically +5.1 V).

It is important to note that the values returned by conventional successive approximation ADCs will not be accurate unless the input voltage has remained substantially constant during the conversion process. Furthermore, where some variation is inevitable, several samples should be taken and averaged.

Analogue output ports are generally much easier to use than their analogue input counterparts. It is usually merely sufficient to output a byte to the appropriate port address. In most cases, analogue output ports will be configured for unipolar operation and, in the case of an 8-bit DAC, a byte value of 00H will result in an output of 0 V while a byte value of FFH will result in a full-scale positive output (typically 5.1 V).

Sensors

Sensors provide a means of inputting information to a process control

176 Interfacing

system. This information relates to external physical conditions such as temperature, position, and pressure. The data returned from the sensors together with control inputs from the operator (where appropriate) will subsequently be used to determine the behaviour of the system.

Any practical industrial process control system will involve the use of a number of devices for sensing a variety of physical parameters. The choice of sensor will be governed by a number of factors including accuracy, resolution, cost and physical size. Table 31 covers the range of sensors and inputs most commonly encountered in industrial process control systems. The list is not exhaustive and details of other types of sensor can be found in most texts devoted to measurement, instrumentation and control systems (see Bibliography).

Table 31

Physical parameter or input	Type of sensor	Notes
Angular position	Resistive rotary position sensor*	Rotary track potentiometer with linear law produces analogue voltage proportional to angular position. Limited angular range. Analogue input port required.
	Shaft encoder*	Encoded disk interposed between optical transmitter and receiver (infra-red LED and photodiode or phototransistor). Usually requires signal conditioning based on operational amplifiers. Digital input port required.
Angular velocity (rotational)	Toothed rotor tachometer	Magnetic pick-up responds to the movement of a toothed ferrous disk. May require signal conditioning (typically an operational amplifier and Schmitt input logic gate). Some sensors contain circuitry to provide TTL compatible outputs. The pulse repetition frequency of the output is proportional to the angular velocity. Digital input port required.
	Tachogenerator	DC generator with linear output characteristic. Analogue output voltage proportional to shaft speed. Requires an analogue input port.
	Shaft encoder*	Encoded disk interposed between optical transmitter and receiver (infra-red LED and photodiode or phototransistor). Requires signal conditioning and some

		additional logic for direction sensing. Digital input port required.
Flow	Rotating vane flow sensor*	Pulse repetition frequency of output is proportional to flow rate. A counter/timer chip can be used to minimize software required. Digital input port required.
Light level	Photocell	Voltage generating device. The analogue output voltage produced is proportional to light level. Analogue input port required.
	Light dependent resistor	Usually connected as part of a potential divider or bridge. An analogue output voltage results from a change of resistance within the sensing element. Analogue input port required.
	Photodiode	Two-terminal device connected as a current source. An analogue output voltage is developed across a series resistor of appropriate value. Analogue input port required.
	Phototransistor	Three-terminal device connected as a current source. An analogue output voltage is developed across a series resistor of appropriate value. Analogue input port required.
Linear position	Resistant linear position sensor*	Linear track potentiometer with linear law produces analogue voltage proportional to linear position. Limited linear range. Analogue input port required.
	Linear variable differential transformer (LVDT)	Miniature transformer with split secondary windings and moving core attached to a plunger. Requires AC excitation and phase sensitive detector. Analogue input port required.
Linear velocity	Magnetic sensor*	Magnetic pick-up responds to movement of a toothed ferrous track. May require signal conditioning (typically using an operational amplifier and Schmitt input logic gates) but some sensors contain circuitry to provide TTL compatible outputs. The pulse repetition frequency of the output is proportional to the linear velocity. Digital input port required.
Liquid level	Float switch*	Simple switch element which operates when a particular level is detected. Digital input port required.
	Capacitive proximity switch*	Switching device which operates when a particular level is detected. Ineffective

		with some liquids. Digital input port required.
	Diffuse scan proximity switch*	Switching device which operates when a particular level is detected. Ineffective with some liquids. Digital input port required.
Operator	Switch or push-button*	Suitable for providing simple on/off control. Available in various styles including keyswitches and foot operated types. Digital input port required.
	DIL switch*	Only suitable for infrequent operation (e.g. setting parameters when re-configuring a system). Digital input port required.
	Keypad*	More cost-effective than using a large number of push-button switches when several options are required. Also suitable for numeric data entry. Keypads fitted with encoders require digital input ports. Unencoded keypads are conventionally configured in the form of a 4 × 4 array and will require at least one digital I/O port.
	Keyboard	Provides the ultimate in data entry (including generation of the full set of ASCII characters). Encoded keyboards are generally easier to use with digital I/O cards than unencoded types which are more suitable for 'memory mapped' I/O. Digital input port required.
	Joystick	Available in digital and analogue forms. The former type is generally based on four microswitches (two for each axis) while the latter is based on conventional rotary potentiometers. Either form is suitable for providing accurate position control. Analogue or digital input ports required, as appropriate.
Pressure	Microswitch pressure sensor	Microswitch fitted with actuator mechanism and range setting springs. Suitable for high pressure applications. Digital input port required.
	Differential pressure/vacuum switch	Microswitch with actuator driven by a diaphragm. May be used to sense differential pressure. Alternatively, one chamber may be evacuated and the sensed pressure applied to the remaining port. Digital input port required.

Sensors 179

	Piezo-resistive pressure sensor	Pressure exerted on diaphragm causes changes of resistance in attached piezo-resistive transducers. Transducers are usually arranged in the form of a four active element bridge which produces an analogue output voltage. Analogue input port required.
Proximity	Microswitch*	Microswitch fitted with actuator. Requires physical contact and small operating force. Digital input port required.
	Reed switch*	Reed switch and permanent magnet actuator. Only effective over short distances. Digital input port required.
	Inductive proximity switch*	Target object modifies magnetic field generated by the sensor. Only suitable for metals (non-ferrous metals with reduced sensitivity). Digital input port required.
	Capacitive proximity switch*	Target object modifies electric field generated by the sensor. Suitable for metals, plastics, wood, and certain powders and liquids. Digital input port required.
	Optical proximity switch*	Available in diffuse and through scan types. Diffuse scan types require reflective targets. Both types employ optical transmitters and receivers (usually infra-red emitting LEDs and photo-diodes or photo-transistors). Digital input port required.
Strain	Resistive strain gauge	Foil type resistive element with polyester backing for attachment to body under stress. Normally connected in full bridge configuration with temperature compensating gauges to provide an analogue output voltage. Analogue input port required.
	Semiconductor strain gauge	Piezo-resistive elements provide greater outputs than comparable resistive foil types. More prone to temperature changes and also inherently non-linear. Analogue input port required.
Temperature	Thermocouple*	Output appears as a small EMF generated by a sensing junction. Requires compensated connecting cables and specialized interface. Analogue input port required.
	Thermistor	Usually connected as part of a potential divider or bridge. An analogue output

180 *Interfacing*

		voltage results from resistance changes within the sensing element. Analogue input port required.
	Semiconductor temperature sensor*	Two-terminal device connected as a current source. An analogue output voltage is developed across a series resistor of appropriate value. Analogue input port required.
Weight	Load cell	Usually comprises four strain gauges attached to a metal frame. This assembly is then loaded and the analogue output voltage produced is proportional to the weight of the load. Requires analogue input port.
Vibration	Electromagnetic vibration sensor	Permanent magnet seismic mass suspended by springs within a cylindrical coil. The frequency and amplitude of the analogue output voltage are respectively proportional to the frequency and amplitude of vibration. Analogue input port required.

* Mentioned later in this chapter.

Interfacing switches and sensors

Sensors can be divided into two main groups according to whether they are active (generating) or passive. Another, arguably more important distinction in the case of bus-based process control systems, is whether they provide digital or analogue outputs. In the former case, one or more digital I/O ports will be required whereas, in the latter case one or more analogue input ports must be provided.

We shall deal first with techniques of interfacing switches and sensors which provide digital outputs (such as switches and proximity detectors) before examining methods used for interfacing sensors which provide analogue outputs. It should be noted that the majority of sensors (of either type) will require some form of signal conditioning circuitry in order to make their outputs acceptable to conventional STE I/O cards.

Switches

Switches can be readily interfaced to STE I/O cards in order to provide manual inputs to the system. Simple toggle and push-button switches are generally available with normally open (NO), normally closed (NC), or changeover contacts. In the latter case, the switch may be configured as either an NO or an NC type, depending upon the connections used.

Toggle, lever, rocker, rotary, slide and push-button types are all commonly available in a variety of styles. Illuminated switches and key switches are also available for special applications. The choice of switch type will obviously depend upon the application and operational environment.

An NO switch or push-button may be interfaced to an STE digital I/O board using nothing more than a single pull-up resistor as shown in Figure 53. The relevant bit of the input port will then return 0 when the switch contacts are closed (i.e., when the switch is operated or when the push-button is depressed). When the switch is inactive, the relevant port bit will return 1.

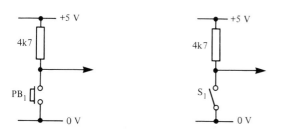

Figure 53 *Interfacing a normally open switch or push-button to a digital input port*

Unfortunately, this simple method of interfacing has a limitation when the state of a switch is regularly changing during program execution. However, a typical application which is unaffected by this problem is that of using one or more PCB mounted switches (for example, a DIL switch package) to configure a system in one of a number of different modes. In such cases, the switches would be set once only and the software would read the state of the switches and use the values returned to configure the system upon reset. Thereafter, the state of the switches would then only be changed in order to modify the operational parameters of the system (for example, when adding additional I/O facilities). A typical DIL switch input interface to a digital input port is shown in Figure 54. Chapter 9 provides a sample application in which the state of such a switch is sensed.

Switch debouncing

As mentioned earlier, the simple circuit of Figure 53 is unsuitable for use when the state of the switch is regularly changing. The reason for this is that the switching action of most switches is far from 'clean' (i.e., the switch contacts make and break several times whenever the switch is

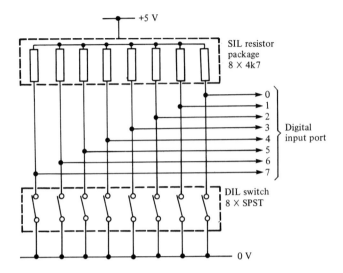

Figure 54 Interfacing a DIL switch input to a digital input port

operated). This may not be a problem when the state of a switch remains static during program execution but it can give rise to serious problems when dealing with, for example, an operator switch bank or keypad.

The contact 'bounce' which occurs when a switch is operated results in rapid making and breaking of the switch until it settles into its new state. Figure 55 shows the waveform generated by the simple switch input circuit of Figure 53 as the contacts close. The spurious states can cause problems if the switch is sensed during the period in which the switch contacts are in motion, and hence steps must be taken to minimize the effects of bounce. This may be achieved by means of additional hardware in the form of a 'debounce' circuit or by including appropriate software delays (of typically 4 ms to 20 ms) so that spurious switching states are ignored. We shall discuss these two techniques separately.

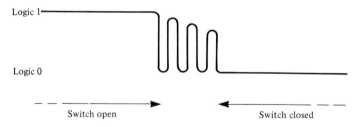

Figure 55 Typical waveform produced by a switch closure

Hardware debouncing

Immunity to transient switching states is generally enhanced by the use of active-low inputs (i.e., a logic 0 state at the input is used to assert the condition required). The debounce circuit shown in Figure 56 is adequate for most toggle, slide and push-button type switches. The value chosen for R_2 must take into account the low-state sink current required by IC_1 (normally 1.6 mA for standard TTL and 400 μA for LS-TTL). R_2 should not be allowed to exceed approximately 470 Ω in order to maintain a valid logic 0 input state. The values quoted generate an approximate 1 ms delay (during which the switch contacts will have settled into their final state). It should be noted that, on power-up, this circuit generates a logic 1 level for approximately 1 ms before the output reverts to a logic 0 in the inactive state. The circuit obeys the following state table.

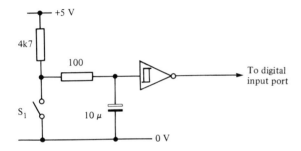

Figure 56 *Simple debounce circuit*

Table 32

Switch condition	Logic output
Closed	1
Open	0

An alternative, but somewhat more complex, switch debouncing arrangement is shown in Figure 57. Here a single-pole double-throw (SPDT) changeover switch is employed. This arrangement has the advantage of providing complementary outputs and it obeys the following state table:

Table 33

Switch condition	Logic output Q	Logic output $\overline{Q}$
Position A	0	1
Position B	1	0

184 Interfacing

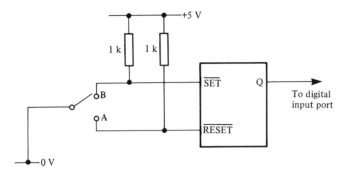

Figure 57 *Debounce circuit based on an RS bistable*

Rather than use an IC RS bistable in the configuration of Figure 57 it is often expedient to make use of 'spare' two-input NAND or NOR gates arranged to form bistables using the circuits shown in Figure 58(a) and (b), respectively. Figure 59 shows a rather neat extension of this theme in the form of a touch operated switch. This arrangement is based on a 4011 CMOS quad two-input NAND gate (though only two gates of the package are actually used in this particular configuration).

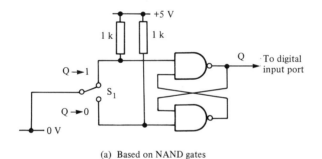

(a) Based on NAND gates

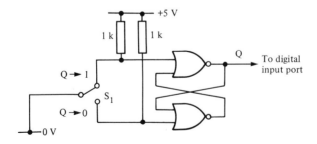

(b) Based on NOR gates

Figure 58 *Alternative switch debounce circuits*
(a) Based on NAND gates   (b) Based on NOR gates

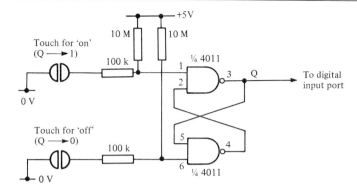

Figure 59 Touch operated switch

Finally, it is sometimes necessary to generate a latching action from an NO push-button switch. Figure 60 shows an arrangement in which a 74LS73 JK bistable is clocked from the output of a debounced switch. Pressing the switch causes the bistable to change state. The bistable then remains in that state until the switch is depressed a second time. If desired, the complementary outputs provided by the bistable may be used to good effect by allowing the $\overline{Q}$ output to drive an LED. This will become illuminated whenever the Q output is high.

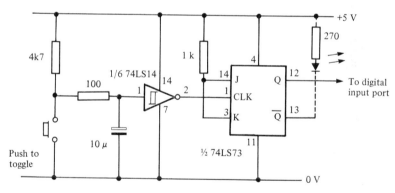

Figure 60 *Latching action switch*

Software debouncing

Software debouncing involves the execution of a delay routine whenever the state of a switch is read. The state of the switch at the start of the delay routine is compared with that at the end. If the same value is returned in both cases, the last value returned is assumed to represent the state of the switch. If the value has changed, the switch is read again. The period of the delay routine is chosen so that it is just greater than the maximum period of contact bounce expected (typically 4 to 10 ms).

A typical software debounce routine is given below:

```
READSW:   CALL  SWITCH      ; Read the switch and
          LD    D,A         ; store the value.
          CALL  SWDELAY     ; Debounce and
          CALL  SWITCH      ; read it again.
          OR    D           ; Has it changed ?
          JR    NZ,READSW   ; Yes, so try again.
          RET               ; No, so return with bit set in A.
SWITCH:   IN    A,(PIOA)    ; Get value from Port A and
          AND   MASK        ; check bit concerned.
          RET               ; Go back...
SWDELAY:  PUSH  AF          ; Preserve the set bit and
          LD    BC,360H     ; delay for approx. 12ms.
SLOOP:    DEC   BC
          LD    A,B
          OR    C
          JR    NZ,SLOOP
          POP   AF
          RET
```

Keypads

Keypads in process control applications vary from simple arrangements

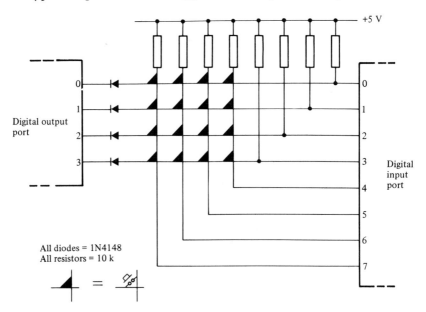

Figure 61 *Typical 4 × 4 matrix keypad interface*

of dedicated push-button switches to arrangements of sixteen-keys (either coded or unencoded) in a standard 4 × 4 matrix. Keycaps may be engraved or fitted with suitable legends. Keypads sealed to IP65 are available as similar units with individually illuminated keys.

Unencoded keypads are invariably interfaced using row and column lines to enable scanning of the keyboard. This arrangement is less demanding in terms of I/O lines than would be the case if the keypad contacts were treated as individual switches. A typical sixteen-key keypad arranged on a 4 × 4 matrix would make use of twelve digital I/O lines though it is possible to use just eight lines of a single port by alternately configuring the port for input and output. A representative arrangement is shown in Figure 61.

Unencoded keypads are generally preferred in high-volume applications where the cost of interfacing hardware has to be balanced at the expense of the extra overhead required by the software involved with scanning the keyboard, detecting and decoding a keypress. In low-volume applications, and where software overheads have to be minimized, the use of a fully encoded keyboard is much to be preferred.

Encoded keypads employ dedicated encoder chips such as the 74C922. This device contains all the necessary logic to interface a 4 × 4 keypad matrix to four lines of a data bus or digital input port. The output is presented in binary coded decimal (BCD) form and an additional signal is provided to indicate that data is available from the keyboard. This active-high data available (DA) output can be used to drive an interrupt line when the keyboard is used in conjunction with a bus processor or may be connected via an open-collector inverter to one of the attention request ($\overline{\text{ATNRQ}}$) lines of the STE bus when the keypad is used in conjunction with a slave card.

The simplified internal arrangement of the 74C922 is shown in Figure 62. The keypad scan may be implemented by the internal clock using an external timing capacitor (C_T) or may be overdriven by an external clock. On-chip pull-up resistors permit keypad switches with contact resistance of up to 50 kΩ. Internal debouncing is provided, the time constant of which is determined by an external capacitor (C_D).

The data available output goes high when a key is depressed and returns to low when a key is released even if another key is depressed. The data available output will return to high to indicate acceptance of the new key after a normal debounce period; this two key rollover is provided between any two switches. An internal register stores the last key pressed even after the key is released.

It should be noted that the LS-TTL compatible outputs of the keypad encoder chip are tristate, thus permitting direct connection to a data bus. Furthermore, the active-low output enable ($\overline{\text{OE}}$) input to the device can be used in a variety of configurations which permit asynchronous data

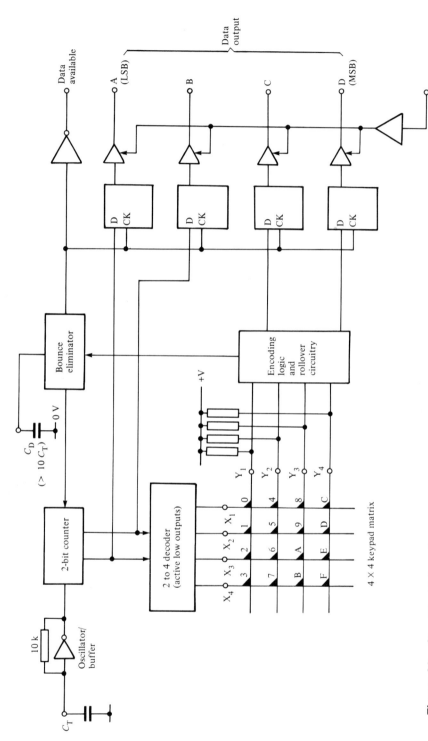

Figure 62 Simplified internal arrangement of the 74C922 keypad encoder

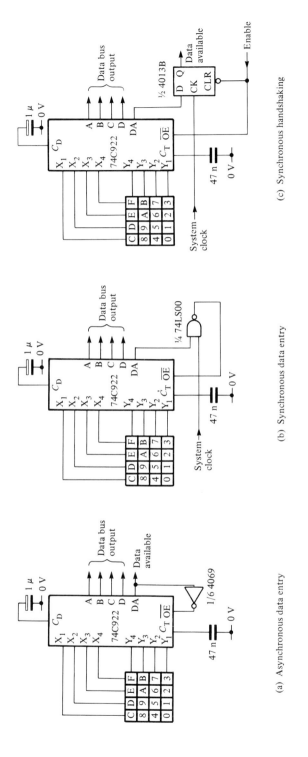

Figure 63 Modes of operation for the 74C922
(a) Asynchronous data entry
(b) Synchronous data entry
(c) Synchronous handshaking

entry as well as synchronous data entry and synchronous handshaking. Figure 63 shows how this can be achieved. Chapter 9 provides further details of using the 74C922 in conjunction with an unencoded keypad in a typical STE application.

Sensors with digital outputs

Sensors which provide digital (rather than analogue) outputs can generally be quite easily interfaced with conventional STE digital I/O cards. However, since the signals generated by such sensors are seldom TTL compatible, it is usually necessary to include additional circuitry between the sensor and input port.

Proximity detectors

Proximity detectors are required in a wide variety of applications; from sensing the presence of an object on a conveyor to detecting whether a machine guard is in place. Simple proximity detectors need consist of nothing more than a microswitch and suitable actuator whereas more complex applications may require the use of inductive or capacitive sensors, or even optical techniques.

Microswitches

A microswitch is a simple electromechanical switch element which requires minimal operating and release force and which exhibits minimal differential travel. Microswitches are normally available in single-pole double-throw (SPDT) configurations and can thus be configured as either normally open (NO) or normally closed (NC).

The principal disadvantage of the humble microswitch is that it not only requires physical contact with the object sensed but also requires a force of typically 40 g to 200 g for successful operation. Most common microswitch types (including the popular V3 and V4 types) can be fitted with a variety of actuator mechanisms. These include lever, roller and standard button types. Metal housed and environmentally sealed microswitches are available for more demanding environments.

Reed switches

Reed switches use an encapsulated reed switch which operates when in the proximity of a permanent magnetic field produced by an actuator magnet. Reed switches are generally available as either normally open

(NO) or changeover types. The latter may, of course, be readily configured for either NO or NC operation. Distances for successful operation (pull-in) of a reed switch are generally within 8 mm to 15 mm (measured between opposite surfaces of the actuator magnet and reed switch assembly). The release range, on the other hand, is generally between 10 mm and 20 mm.

Inductive proximity detectors

Inductive proximity detectors may be used for sensing the presence of

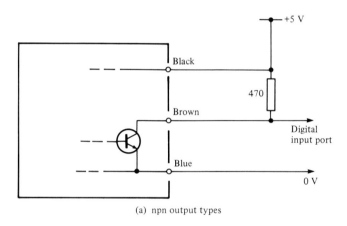

(a) npn output types

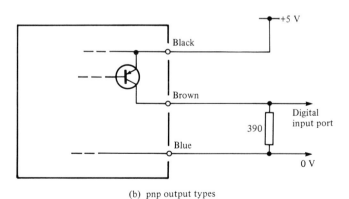

(b) pnp output types

Figure 64 *Interfacing inductive proximity sensors*
(a) n-p-n output types
(b) p-n-p output types

metal objects without the need for any physical contact between the object and the sensor. Inductive proximity switches can be used to detect both ferrous and non-ferrous metals (the latter with reduced sensitivity). Hence metals such as aluminium, copper, brass and steel can all be detected. Typical sensing distances for milk steel targets range from 1 mm for an object having dimensions 4 mm × 4 mm × 1 mm to 15 mm for an object measuring 45 mm × 45 mm × 1 mm. Note that sensitivity is reduced to typically 35 per cent of the above for non-ferrous metals such as aluminium, brass and copper.

Inductive proximity detectors are available with either n-p-n or p-n-p outputs (as shown in Figure 64(a) and (b)). An n-p-n-type will return a logic 0 (low) when a target is detected while a p-n-p-type will return logic 1 (high) in similar circumstances. When selecting a transducer for use with conventional I/O cards, it is advisable to choose a device which operates from a +5 V supply as this obviates the need for level shifting within the interface. A further consideration with such devices is the maximum speed at which they can operate. This is typically 2 kHz (i.e., 2000 pulses per second) but note that some devices are very much slower.

Capacitive proximity detectors

Capacitive proximity detectors provide an alternative solution to the use of inductive sensors. Unfortunately, such devices are also limited in their speed of response (typically 250 Hz maximum) and often require supply voltages in excess of the conventional +5 V associated with TTL signals. Capacitive proximity sensors will, however, detect the presence of materials such as cardboard, wood and plastics as well as certain powders and liquids. Typical sensing distances range from 20 mm for metals to 4 mm for cardboard. As with inductive proximity sensors, the sensitivity of the detector is proportional to target size. A typical interface circuit for a DC powered capacitive proximity detector is shown in Figure 65. This circuit provides a logic 0 (low) when a target is detected.

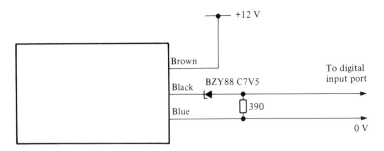

Figure 65 *Interface circuit for a typical capacitive proximity sensor*

Optical proximity detectors

Optical proximity detectors generally offer increased sensing ranges in comparison with both capacitive and inductive types. Optical proximity sensors are available in two basic forms: diffuse scan and through scan types. The former types rely on the target surface returning a proportion of the modulated light emitted by an optical transmitter which is mounted in the same enclosure as the receiver. In such an arrangement, a reflective target may be detected by the presence of a received signal. Through scan types, on the other hand, employ a separate transmitter and receiver and operate on the principle of the interrupted light beam (i.e., the target is detected by the absence of received light). Typical ranges vary from about 100 mm to 300 mm for diffuse scan sensors with plain white surfaces to up to 15 m for through scan sensors with opaque targets.

Proprietary sensor units are generally rather slow in operation and, for applications which involve rapid motion (such as counting shaft speeds) faster sensors should be employed. Here, a simple optical sensor (comprising an unmodulated infra-red emitting LED and photodiode) may be employed. Such devices are readily available in a variety of packages including miniature diffuse scan types and slotted through scan units. Figure 66 shows the circuitry required to interface such a device to a typical digital input port.

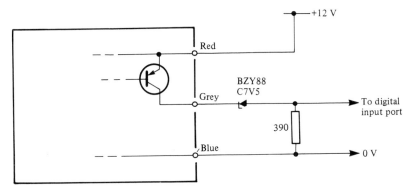

Figure 66 *Interface circuit for an optical proximity sensor*

Position transducers

Position transducers can be used to provide an accurate indication of the position of an object and are available in a variety of forms (including linear and rotary types). Linear position sensors use linear law potentiometer elements (of typically 5 kΩ) and offer strokes of typically 10 mm or

100 mm. Rotary position sensors are also available. These provide indications over typically 105° and use linear law potentiometer elements similar to those found in conventional rotary potentiometer controls. A typical value for the resistive element is again 5 kΩ.

The output of linear and rotary position sensors is usually made available as an analogue voltage and a typical arrangement is shown in Figure 67. Note that the analogue input port should have a high impedance (say 500 kΩ, or more) in order to avoid non-linearity caused by loading of the sensing potentiometer.

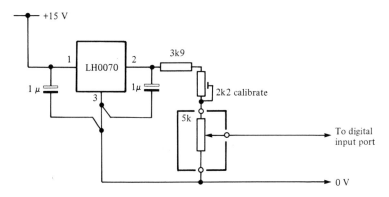

Figure 67 *Interface circuit for a resistive position transducer (either linear or rotary type)*

Shaft encoders

Shaft encoders can be used for sensing both rotary position and shaft speed. A typical shaft encoder produces 100 pulses per revolution and can thus provide a resolution of better than 1°. Such a device generally produces two phase-shifted outputs (to enable detection of direction of rotation) plus a third synchronizing pulse output (one pulse per revolution).

Shaft encoders are generally supplied in kit form comprising an encoder module, slotted disk and hub. The encoder module usually contains three infra-red emitting LEDs and three matching photodetectors. The slotted disk is bonded to the hub ring which is, in turn, fitted to the rotating shaft. The encoder module is then mounted so that the disk is interposed between the LEDs and photodetectors.

The outputs of the encoder module are sinusoidal (as shown in Figure 68) and these must be converted to TTL compatible input pulses in order to interface with a standard digital input port. For simple speed sensing applications, a typical input stage based on an operational comparator and low-pass filter is shown in Figure 69.

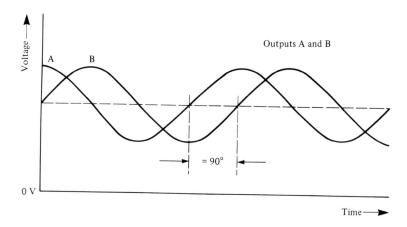

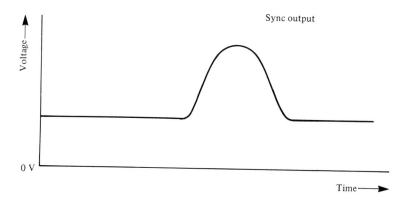

Figure 68 *Output waveforms produced by a typical shaft encoder*

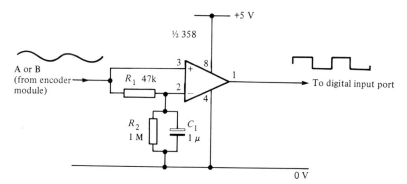

Figure 69 *Shaft encoder signal conditioning for measurement of rotational speed*

Unfortunately, the simple circuit of Figure 69 is ineffective at very low frequencies and for stationary position indication. In such cases, the circuit shown in Figure 70 may be employed. Here, the potentiometer must be adjusted so that the potential at the inverting input of the comparator is equal to that present at the non-inverting input. In this condition, the comparator produces a near 50 per cent duty cycle.

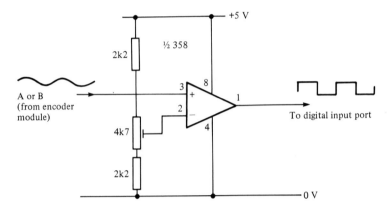

Figure 70 *Shaft encoder signal conditioning for low speed applications and position sensing*

A further refinement is that of providing an output which indicates the sense of rotation (i.e., clockwise or anticlockwise). This may be achieved with the aid of some additional logic and a single JK bistable element as shown in Figure 71. The Q output of the bistable goes high (logic 1) for clockwise rotation and low (logic 0) for anticlockwise rotation. Figure 72 shows typical waveforms for the logic shown in Figure 71.

Fluid sensors

A number of specialized sensors are available for use with fluids. These sensors include float switches (both horizontal and vertical types) and flow sensors. These latter devices incorporate a rotating vane and are suitable for use with flow rates ranging from 3 litres/hour to 500 litres/hour. Typical outputs from 24 Hz at 10 litres/hour to 52 Hz at 20 litres/hour.

Optically isolated inputs

In a number of applications, it may be necessary to provide a high degree

Optically isolated inputs 197

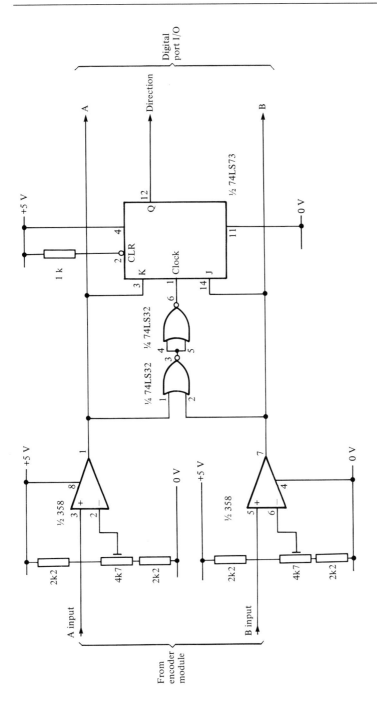

Figure 71 Additional logic required to provide direction sensing

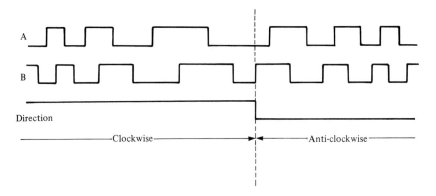

Figure 72 *Typical waveforms produced by the circuit of Figure 71*

of electrical isolation between the source of a digital signal and its eventual connection to a digital input port. Such isolation can be achieved with the aid of an opto-isolator. These units comprise an optically coupled infra-red emitting LED and photodetector encapsulated in DIL package. The photodetector may take various forms including a photodiode, phototransistor and photo-Darlington. Typical isolation voltages provided by such devices range from 500 V to 3 kV and switching rates may be up to 300 kHz or so. High voltage opto-isolators are available which will work reliably at voltages of up to 10 kV.

A typical single-channel optically isolated input arrangement is depicted in Figure 73. The external diode protects the infra-red emitting LED from inadvertent reversal of the input polarity and the value of the series resistor should be selected from Table 34.

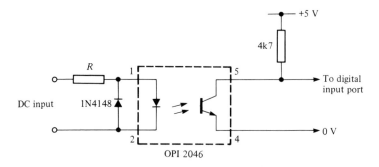

Figure 73 *Optically isolated digital input*

Table 34

Input voltage range (V)	Series resistor, R (Ω)
3 to 4	330
4 to 5	560
5 to 6	680
6 to 8	1 k
8 to 11	1.5 k
11 to 15	2.2 k

The optically-isolated input stage can be extended for monitoring AC voltages as shown in Figure 74. This arrangement is suitable for AC inputs of up to 240 V 50 Hz and may be used to sense the presence or absence of a mains supply. The circuit is, however, relatively slow in operation and fails to meet the specification for an AC power failure signal (via ATNRQO*) within an STEbus system.

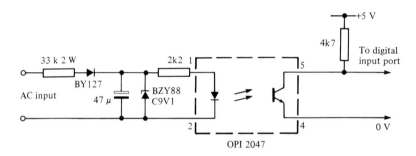

Figure 74 *Modification of Figure 71 to permit AC input*

Sensors with analogue outputs

Having dealt with a number of common sensors which provide digital outputs, we shall now turn our attention to a range of transducers which provide analogue outputs. These outputs may manifest themselves as changes in EMF, resistance, or current and, in the latter cases it will usually be necessary to incorporate additional signal conditioning circuitry so that an analogue input voltage can be provided for use with a standard STE analogue input card.

Semiconductor temperature sensors

Semiconductor temperature sensors are ideal for a wide range of temperature sensing applications. The popular 590kH semiconductor temperature sensor, for example, produces an output current which is proportional to absolute temperature and which increases at the rate of 1 µA/°K. The characteristic of the device is illustrated in Figure 75.

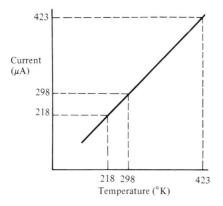

Figure 75 *Characteristic of the 590 kH semiconductor temperature sensor*

The 590kH is laser trimmed to produce a current of 298.2 µA (±2.5 µA) at a temperature of 298.2°C (i.e., 25°C). A typical interface between the 590 kHz and an analogue input port is shown in Figure 76.

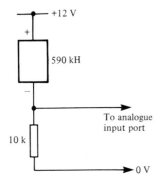

Figure 76 *Typical input interface for the 590 kH semiconductor temperature sensor*

Thermocouples

Thermocouples comprise a junction of dissimilar metals which generates an EMF proportional to the temperature differential which exists between the measuring junction and a reference junction. Since the measuring junction is usually at a greater temperature than that of the reference junction, it is sometimes referred to as the *hot junction*. Furthermore, the reference junction (*cold junction*) is often omitted in which case the sensing junction is simply terminated at the signal conditioning board. This board is usually maintained at, or near, normal room temperatures.

Thermocouples are suitable for use over a very wide range of temperatures (from $-100°C$ to $+1100°C$). Industry standard 'type K' thermocouples comprise a positive arm (conventionally coloured brown) manufactured from nickel/chromium alloy while the negative arm (conventionally coloured blue) is manufactured from nickel/aluminium.

The characteristic of a type K thermocouple is defined in BS 4937: Part 4: 1973 *International Thermocouple Reference Tables*. This standard gives tables of EMF versus temperature over the range $0°C$ to $1100°C$. In order to minimize errors, it is usually necessary to connect thermocouples to appropriate signal conditioning using compensated cables and matching connectors. Such cables and connectors are available from a variety of suppliers and are usually specified for use with type K thermocouples.

Where thermocouples are to be used as sensors in conjunction with STE systems, proprietary signal conditioning cards are available. These cards incorporate cable terminators and provide cold junction compensation as well as low-pass filtering to reduce the effects of 50 Hz noise induced in the thermocouple cables. The signal conditioning boards are then used in conjunction with one or more multichannel analogue input ports.

Threshold detection with analogue output transducers

Analogue sensors are sometimes used in situations where it is only necessary to respond to a predetermined threshold value. In effect, a two-state digital output is required. In such cases a simple 1-bit analogue to digital converter based on a comparator can be used. Such an arrangement is, of course, very much simpler and more cost-effective than making use of a conventional analogue input port.

Simple threshold detectors for light level and temperature are shown in Figures 77 to 79. These circuits produce TTL compatible outputs suitable for direct connection to a digital input port.

Figure 77 shows a light level threshold detector based on a comparator and light dependent resistor (LDR). This arrangement generates a logic 0 input whenever the light level exceeds the threshold setting, and vice versa. Figure 78 shows how light level can be sensed using a photodiode. This circuit behaves in the same manner as the LDR equivalent but readers should be aware that circuit achieves peak sensitivity in the near infra-red region.

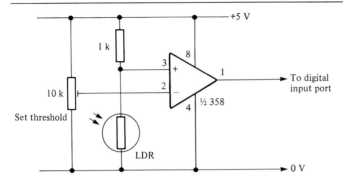

Figure 77 *Light level threshold detector based on a light dependent resistor (LDR)*

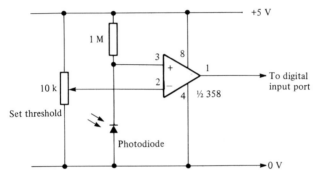

Figure 78 *Light level threshold detector based on a photodiode*

Figure 79 shows how temperature thresholds can be sensed using the 590 kH sensor described earlier. This arrangement generates a logic 0 input whenever the temperature level exceeds the threshold setting, and vice versa.

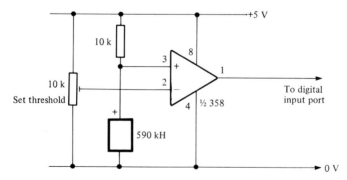

Figure 79 *Temperature threshold detector based on a 590 kH semiconductor temperature sensor*

AC sensing

Finally, Figure 80 shows how an external AC source can be coupled to an input port. This arrangement produces TTL compatible input pulses having 50 per cent duty cycle. The circuit requires an input of greater than 10 mV r.m.s. for frequencies up to 10 kHz and greater than 10 mV r.m.s. for frequencies up to 100 kHz.

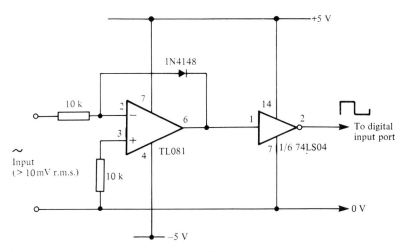

Figure 80 *Interface circuit to permit AC sensing*

The obvious application the arrangement shown in Figure 80 is the detection of audio frequency signals but, with its input derived from the low voltage secondary of a mains transformer (via a 10:1 potential divider), it can also function as a mains failure detector.

Output devices

Having dealt at some length with input sensors, we shall now focus our attention on output devices and the methods used for interfacing them. The STE bus can readily be configured to work with a variety of different output transducers including actuators, alarms, heaters, lamps, motors and relays. Ready-built output drivers are available for several types of load including relays and stepper motors. Many applications will, however, require custom-built circuitry in order to interface the necessary output devices.

Status and warning indications

Indications based on light emitting diodes (LEDs) are inherently more reliable than small filament lamps and also consume considerably less power. They are thus ideal for providing visual status and warning displays. LEDs are available in a variety of styles and colours and 'high brightness' types can be employed where high intensity displays are required.

A typical red LED requires a current of around 10 mA to provide a reasonably bright display and such a device may be directly driven from a buffered digital output port. Different connections are used depending upon whether the LED is to be illuminated for a logic 0 or logic 1 state. Several possibilities are shown in Figure 81.

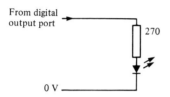

(a) Logic 1 to illuminate the LED

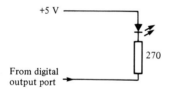

(b) Logic 0 to illuminate the LED

Figure 81 Driving an LED for a buffered digital I/O port
(a) Logic 1 to illuminate the LED
(b) Logic 0 to illuminate the LED

Where a buffered output port is not available, an auxiliary transistor may be employed, as shown in Figure 82. The LED will operate when the output from the STE I/O card is taken to logic 1 and the operating current should be approximately 15 mA (thereby providing a brighter display than the arrangements previously described). The value of series resistor will be dependent upon the supply voltage and should be selected from Table 35.

Driving LCD displays

A number of process control applications require the generation of status messages and operator prompts. These can be easily produced using a

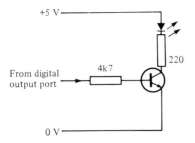

Figure 82 *Using an auxiliary transistor to drive an LED*

Table 35

Supply voltage (V)	R_2 (Ω)
3 to 4	100
4 to 5	150
5 to 8	220
8 to 12	470
10 to 15	820

conventional alphanumeric dot matrix LCD display. Such displays are commonly available in a variety of formats ranging from 16 characters × 1 line to 40 characters × 4 lines. LCD displays are invariably fitted with the necessary logic (often in the form of a CMOS microprocessor) to interface directly with a bus. Furthermore, such modules can display the full ASCII character set as well as user-defined symbols.

In STEbus I/O applications, LCD displays will generally require the services of a byte-wide I/O port. However, since modules can be directly interfaced with a system bus, a number of proprietary STE cards incorporate provision for the attachment of an LCD display. A typical application involving the use of an LCD display is discussed in Chapter 9.

Driving medium and high-current loads

Owing to the limited output current and voltage capability of most standard STE I/O ports, external circuitry will normally be required to drive anything other than the most modest of loads. Figure 83 shows some typical arrangements for operating various types of medium and high current load. Figure 83(a) shows how an n-p-n transistor can be used to operate a low-power relay. Where the relay requires an appreciable operating current (say, 150 mA or more) a plastic encapsulated Darlington power transistor should be used as shown in Figure 83(b). Alternatively, a VMOS power FET may be preferred, as shown in Figure

206 Interfacing

83(c). Such devices offer very low values of 'on' resistance coupled with a very high 'off' resistance. Furthermore, unlike conventional bipolar transistors, such devices impose a negligible load on the I/O port. Figure 83(d) shows a filament lamp driver based on a plastic Darlington power transistor. This circuit will drive lamps rated at up to 24 V, 500 mA.

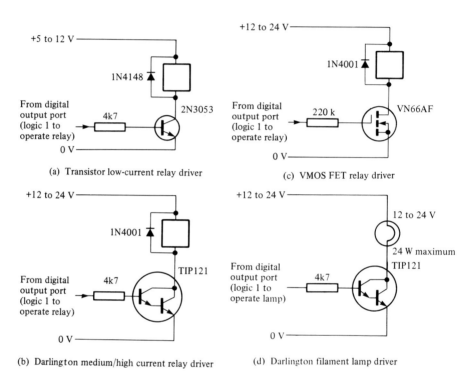

Figure 83 *Typical medium and high-current driver circuits*
(a) Transistor low-current relay driver
(b) Darlington medium/high current relay driver
(c) VMOS FET relay driver
(d) Darlington filament lamp driver

Audible outputs

Where simple audible warnings are required, miniature piezo-electric transducers may be used. Such devices operate at low voltages (typically in the range 3 V to 15 V) and can be interfaced with the aid of a buffer, open-collector logic gate, or transistor. Figure 84(a) to (c) shows typical interface circuits which produce an audible output when the port output line is at logic 1.

Where a pulsed rather than continuous audible alarm is required, a circuit of the type shown in Figure 85 can be employed. This circuit is

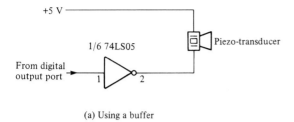

(a) Using a buffer

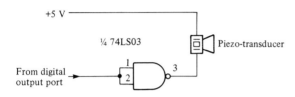

(b) Using an open-collector logic gate

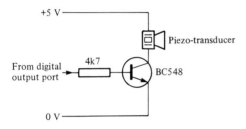

(c) Using a transistor

Figure 84 *Audible output driver*
(a) Using a buffer
(b) Using an open-collector logic gate
(c) Using a transistor

based on a standard 555 timer operating in astable mode and operates at approximately 1 Hz. A logic 1 from the port output enables the 555 and activates the pulsed audio output.

Finally, the circuit shown in Figure 86 can be used where a conventional moving coil loudspeaker is to be used in preference to a piezo-electric transducer. This circuit is again based on the 555 timer and provides a continuous output at approximately 1 kHz whenever the port output is at logic 1.

208 Interfacing

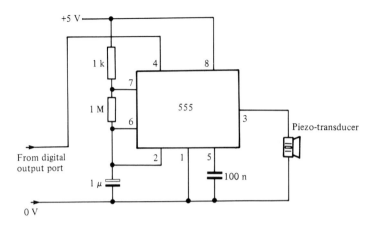

Figure 85 *Pulsed audible alarm based on a 555 astable*

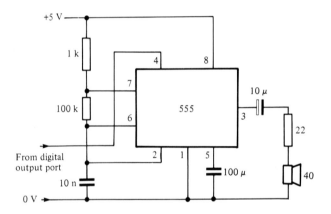

Figure 86 *Audible alarm with output to a moving coil loudspeaker*

DC motors

Circuit arrangements used for driving DC motors generally follow the same lines as those described earlier for use with relays. As an example, the circuit shown in Figure 87 uses a VMOS FET to drive a low-voltage DC motor. This circuit is suitable for use with DC motors rated at up to 12 V with stalled currents of less than 1.5 A. A logic 1 from the output port operates the motor.

Byte-wide output drivers

Where a number of output loads are to be driven from the same port, it is often expedient to make use of a dedicated octal driver chip rather than

Output devices 209

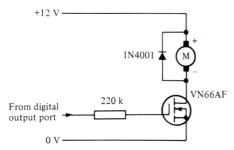

Figure 87 *VMOS FET DC motor driver*

use eight individual driver circuits based on discrete components. Fortunately a number of octal drivers are available and these invariably have TTL compatible inputs which makes them suitable for direct connection to an output port.

A simple, general purpose byte-wide output driver can be based around a dedicated octal latch/driver of which the UCN5801A is a typical example. This device is directly bus compatible but may also be used in

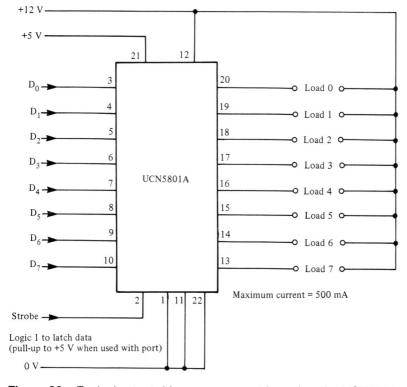

Figure 88 *Typical output driver arrangement based on the UCN5801*

conjunction with a conventional parallel I/O port. The UCN5801A has separate CLEAR, STROBE and output ENABLE control lines coupled to eight bipolar Darlington driver transistors. This configuration provides an extremely low power latch with a very high output current capability.

The eight outputs of the UCN5801A are all open-collector, the positive supply voltage for which may be anything up to 50 V. Each Darlington output device is rated at 500 mA maximum. However, if that should prove insufficient for a particular application then several output lines may be paralled together subject, of course, to the limits imposed by the rated load current of the high voltage supply.

Figure 88 shows a typical arrangement of the UCN5801 in which the load voltage supply is +12 V. The state of the bus is latched to the output whenever the STROBE input is taken high however, when used in connection with a conventional I/O device, this line can be tied high. A logic 0 present on a particular data line will turn the corresponding Darlington output device 'off' while a logic 1 will turn it 'on'. It should also be noted that the output stages are protected against the effects of an inductive load by means of internal diodes. These are commoned at pin-12 and this point should thus be returned to the positive supply. Chapter 9 describes an application involving the use of the UCN5801 as an output driver.

Driving mains connected loads

Control systems are often used in conjunction with mains connected loads. Modern solid-state relays (SSRs) offer superior performance and reliability when compared with conventional relays in such applications. SSRs are available in a variety of encapsulations (including DIL, SIL, flatpack, and plug-in octal) and may be rated for RMS currents between 1 A and 40 A.

In order to provide a high degree of isolation between input and output, SSRs are optically coupled. Such devices require minimal input currents (typically 5 mA, or so, when driven from 5 V) and they can thus be readily interfaced with TTL devices. Figure 89(a) to (c) shows how an SSR can be driven from buffered and unbuffered I/O ports. It is important to note that, when an inductive load is to be controlled, a *snubber* network should be fitted, as shown in Figure 90.

Driving solenoids and solenoid operated valves

Solenoids and solenoid operated valves are generally available with coils rated for 110 V/240 V AC or 12 V/24 V DC operation. The circuitry for interfacing solenoids will thus depend on whether the unit is rated for AC

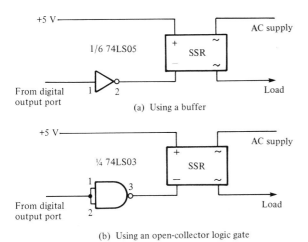

Figure 89 *Interface circuits for driving solid state relays*
(a) Using an auxiliary buffer stage
(b) Using an open-collector logic gate

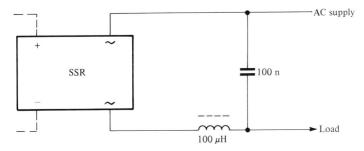

Figure 90 Using a 'snubber' network with an inductive load

or DC operation. In the case of AC operated units, a suitably rated SSR should be employed (see Figure 89(a) and (b)) while, in the case of DC operated solenoids, interface circuitry should be identical to that employed with medium/high current relays (see Figure 83(b) and (c)).

Driving stepper motors

The complex task of interfacing a stepper motor to an STEbus system can be much simplified by using a dedicated stepper motor driver chip such as the SAA1027. This device includes all necessary logic to drive a stepper motor as well as output drivers for each of the four phases. The chip operates from a nominal +12 V supply rail but its inputs are not directly TTL compatible and thus transistor or open-collector drivers will normally be required.

Figure 91 shows a typical stepper motor interface based on the SAA1027. The motor is a commonly available four-phase two-stator type having the following characteristics:

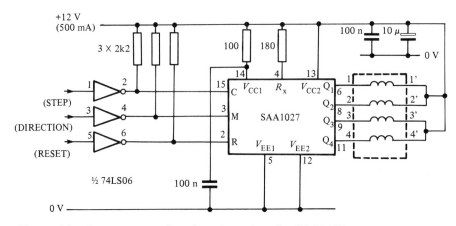

Figure 91 Stepper motor interface based on the SAA1027

Supply voltage	12 V
Resistance per phase	47 Ω
Inductance per phase	400 mH
Maximum working torque	50 mNm
Step rotation	7.5°/step

The stepper motor interface requires only three port output lines which operate on the following basis:

1. The $\overline{\text{STEP}}$ input is pulsed low to produce a step rotation.
2. The DIRECTION input determines the sense of rotation. A low on the DIRECTION input selects clockwise rotation. Conversely, a high on the DIRECTION input selects anticlockwise rotation.
3. The $\overline{\text{RESET}}$ input can be taken low to reset the driver.

The software routines for driving the stepper motor are quite straightforward and Chapter 9 describes an application in which a stepper motor is interfaced to the STE I/O bus.

9
STEbus applications

The STEbus is at home in a huge variety of process control applications ranging from simple stand alone machine controllers to fully integrated factory control systems. This chapter aims to provide readers with an introduction to the procedure for selecting and specifying hardware and software for bus-based industrial process control systems. In addition,

Figure 92 *A forty-channel digital I/O card which also provides four serial channels (courtesy of DSP Design Ltd)*

two representative applications of STEbus systems are discussed at some length.

STE products

The range of STE boards currently available from over twenty manufacturers includes:

- Analogue I/O cards with up to sixteen analogue inputs and up to four buffered analogue outputs (converters generally offer either 8- or 12-bit resolution).
- Digital I/O cards with up to forty buffered I/O lines.
- Disk and tape controllers providing control for up to four 3.5, 5.25 or 8 in floppy disk drives or SCSI compatible Winchester drives or tape streamers (Figure 93).

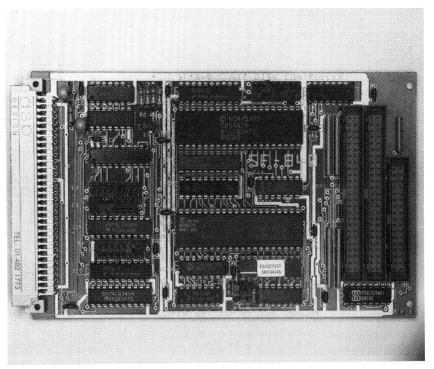

Figure 93   *Floppy and SCSI disk controller card (courtesy of DSP Design Ltd)*

- EPROM programmers (for most popular EPROM devices).
- IEEE-488 GPIB interface cards.
- Memory cards (with up to 512 kilobytes of dynamic RAM) (Figure 94).

Figure 94 *Memory card incorporating ROM and CMOS static RAM (courtesy of DSP Design Ltd)*

- Modem cards (to permit transfer of data via conventional telephone lines).
- Network interface cards.
- Operator interface cards (allowing keypads, keyboards and LCD displays to be interfaced with STE bus systems).
- Processor cards based on the most popular 8- and 16-bit microprocessors (including the 64180, 65C02, 65816, 6809, 68008, 80C88, 80188, and Z80) (Figures 95 and 96).
- Prototyping cards (these comprise a bus slave I/O interface, have link selectable base addresses, and provide the user with an area for soldering components fitted into a 0.1 in matrix of plated through holes).
- Real-time clock cards (often also fitted with up to 24 kilobytes of battery backed CMOS static RAM).
- Serial communications cards with up to four RS-232C, RS-422 or RS-423 serial ports.
- Stepper motor controllers and shaft encoders.
- Video cards (which permit text and graphic displays when used with monochrome or colour monitors).
- Watchdog controllers for system monitoring and bus arbitration.

216 STEbus applications

Figure 95 *Processor card with forty parallel I/O lines. The processor is an HD64180 (enhanced Z80) (courtesy of DSP Design Ltd)*

Figure 96 *Processor card with on-board battery packed RAM. The processor is an HD64180 (enhanced Z80) (courtesy of DSP Design Ltd)*

In addition, the system builder is able to select from a large range of signal conditioning cards in standard Eurocard format. These provide the necessary interfacing circuitry for a wide range of popular sensors and output devices. It is thus eminently possible to construct an STE-based process control system simply by selecting 'off-the-shelf' modules; only when dealing with very specialized applications is it necessary to manufacture one's own slave I/O cards or signal conditioning boards. The range of STE products currently available can be found in the *STE Bus Product Guide* (see Appendix 3) but several component distributors maintain stocks of the most popular bus and signal conditioning boards.

Figure 97 *A forty-channel optically isolated digital input signal conditioning card (courtesy of DSP Design Ltd)*

In conjunction with appropriate software, the STEbus is also ideal for use as a low-cost development system; allowing software and hardware to be developed for specialized target applications. Complete STE-based development systems are available from several manufacturers. However, in order to meet the user's specific a simple development system can quite easily be assembled from off-the-shelf modules and software.

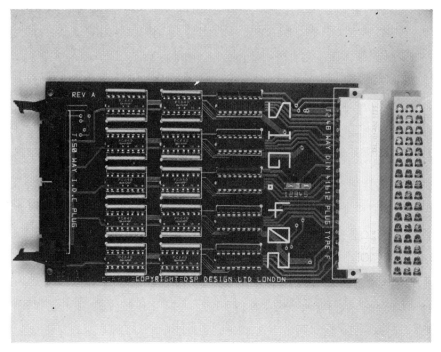

Figure 98 *A forty-channel optically isolated digital output signal conditioning card (courtesy of DSP Design Ltd)*

Communications and networking

Unless one is dealing with totally stand-alone systems, most modern STE applications will require some capability for communications. This can often be provided by incorporating processors or I/O boards fitted with a standard RS-232C serial interface. Such an interface will be adequate for short distance communications on a 'one to one' basis (for example to permit data transfer from an STE system to a nearby personal computer) however, when a number of STE systems are present in a given area, a need may arise for some form of multi-drop networked system. This can be realized using a standard such as that exemplified by Intel's BITBUS.

BITBUS is a serial data bus-based on the RS-485 physical and electrical interface standard (RS-485 is a multidrop version of RS-422) and the datalink protocol employed is a subset of SDLC/HDLC. BITBUS is complementary to manufacturing automation protocol (MAP) which has gained widespread recognition as the industrial standard for the upper level of factory data communications. At the machine and process level, however, where time critical data from sensors, actuators and alarms has

Communications and networking 219

to be transmitted, the response time of MAP, though guaranteed, is inadequate. BITBUS, on the other hand, is well suited to the transfer of short *field data* messages.

BITBUS is configured as a single-master, multi-slave network and operates in one of two modes, synchronous and self-clocked. Synchronous operation permits speeds of up to 2.4 megabit/s but requires twin twisted-pair cables and is restricted to transmission over distances less than 300 m. Furthermore, since repeaters cannot be used in this mode, a maximum of 31 nodes is possible. Self-clocked mode, on the other hand, requires only single pair cable, can operate at either 62.5 or 375 kilobit/s

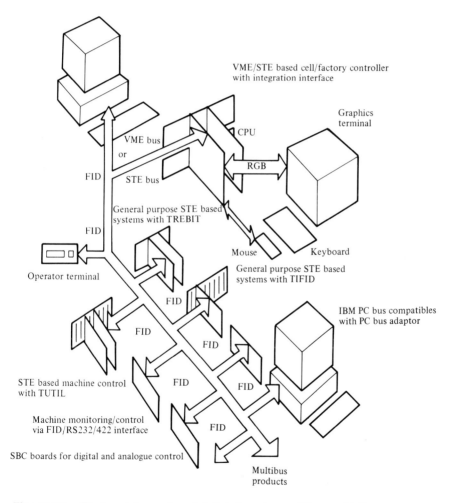

Figure 99 *Factory information datalink (courtesy of Trenan Ltd)*

and, with repeaters, can cater for up to 250 nodes at distances not exceeding 13 km.

Interfacing with BITBUS is usually made possible with the use of an Intel 8044 micro-controller which implements the BITBUS protocol using an on-chip SDLC controller and ROM-based firmware. An interface of this type may be incorporated within a processor card or may be provided as part of an auxiliary communications interface. The Trenan factory information datalink (FID), shown in Figure 99, exemplifies the use of networked controllers in a modern integrated manufacturing system.

Specifying STEbus hardware and software

When specifying hardware and software to be used in a given STEbus application, it is essential to adopt a 'top-down' approach. An important first stage in this process is that of defining the overall aims of the system before attempting to formalize a detailed specification. The aims should be agreed with the end-user and should be reviewed within the constraints of available budget and time. Specifications should then be formalized in sufficient detail for the performance of the system to be measured against them and should include such items as input and output parameters, response time, accuracy and resolution.

Having set out a detailed specification, it will be possible to identify the main hardware elements of the system as well as the types of sensor and output device required (see Chapter 8). The following checklist, arranged under six major headings, should assist in this process:

1 Performance specification.
   What are the parameters of the system?
   What accuracy and resolution is required?
   What aspects of the process are time critical?
   What environment will the equipment be used in?
   What special contingencies should be planned for?
   What degree of fault-tolerance is required?
2 I/O devices.
   What sensors will be required?
   What output devices will be required?
   What I/O and signal conditioning boards will be required?
   Will it be necessary to provide high-current or high-voltage drivers?
   Should any of the inputs or outputs be optically isolated?
3 Displays and operator inputs.
   What expertise can be assumed on the part of the operator?
   What alarms and status displays should be provided?

What inputs are required from the operator?
What provision for resetting the system should be incorporated?
4. Program/data storage.
What storage medium and format is to be employed?
How much storage space will be required for the operating system and/or control program?
How much storage space will be required for data?
How often will the control program need updating?
Will stored data be regularly updated during program execution?
What degree of data security must be achieved?
5. Communications.
What existing communications standards are employed by the end-user?
Will a standard serial data link based on RS-232 be sufficient or will a faster, low-impedance serial data communications standard be needed?
What data rates will be required?
What distances are involved?
Will it be necessary to interface with automatic test equipment?
Will a networking capability be required?
6. Expansion.
What additional facilities are envisaged by the end-user?
What additional facilities could be easily incorporated?
Will expansion necessitate additional hardware, additional software, or both?
What provision should be made for accommodating additional hardware?

Hardware design

Start by identifying the principal elements of the system including card frame, power supply, etc. Then itemize the input devices (such as keypads, switches and sensors), and output devices (such as motors, actuators and displays). This process may be aided by developing a diagram of the system showing the complete hardware configuration and the links which exist between the elements. This diagram will subsequently be refined and modified but initially will serve as a definition of the hardware components of the system.

Having identified the inputs required, a suitable sensor or input device should be selected for each input (see Chapter 8). It should then be possible to specify any specialized input signal conditioning required with reference to the manufacturer's specification for the sensor concerned. Input signal conditioning should then be added to the system diagram mentioned earlier.

Next, a suitable driver or output interface should be selected for each output device present (see Chapter 8). Any additional output signal conditioning required should also be specified and incorporated in the system diagram.

Software design

Software design should mirror the top-down approach adopted in relation to the system as a whole. At an early stage, it will be necessary to give some consideration to the overall structure of the program and identify each of the major functional elements of the software and their relationship within the system as a whole. It is important to consider the constraints of the system imposed by time critical processes and hardware limitations (such as the size of available memory). Furthermore, routines to cope with input and output may require special techniques (for example, specialized assembly language routines).

The software should be designed so that it is easy to maintain, modify and extend. Furthermore, the programmer should use or adapt modules ported from other programs. These modules will already have been proven and their use should be instrumental in minimizing development time.

When developing software, it is advisable to employ only *simple logic* (i.e., that which has been tried and understood). The temptation to produce untried and overcomplicated code should be avoided. Simple methods will usually produce code which is easy to maintain and debug, even if the code produced requires more memory space or executes more slowly. If the process is time critical or memory space is at a premium then code can later be refined and optimized. It is also important to consider all eventualities which may arise, not just those typical of normal operation. The following are particularly important:

- Will the system initialize itself in a safe state – will there be momentary unwanted outputs during start-up?
- What will happen if the user defaults an input or if an input sensor becomes disconnected?
- What will happen if the power fails – will the system shut-down safely?
- What input validation checks are required – what steps should be taken if an 'out of range' input is detected?

Applications

The remainder of this chapter provides details of two representative STE bus applications. Neither of these applications is particularly novel but they both address problems which are typical of those which face the process control engineer. The applications have been chosen to illustrate contrasting aspects of design and, while it would be impossible to describe either of these applications in their entirety, they should provide readers with a feel for the process of designing and implementing a bus-based control system.

A profile digitizer

A profile digitizer can be used to develop a digital image of an item. This application shows how the STE bus can be used in a data acquisition system where the bus is also used to control external hardware.

Specification

The end-user requires a device capable of automatically scanning a test piece in order to produce a stored digitized profile. Digitized profiles are

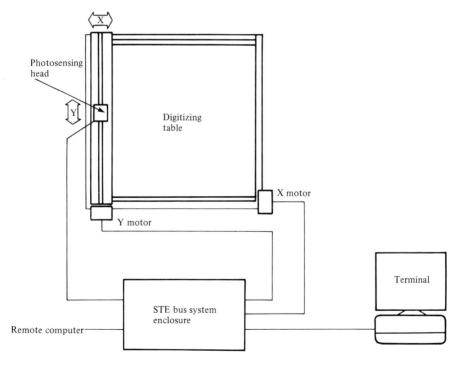

Figure 100 *Basic arrangement of the profile digitizer*

to be stored on disk and may later be transmitted, via an RS-232C data link, to an external computer system. The scanning process is not particularly time critical but should be carried out in as short a time as possible. Two scales should be provided (×1 and ×2) and, in each case, the digitized profile should comprise 1024 × 1024 points in each axis.

The digitizer should comprise a table on to which the test piece is placed together with an optical sensing head which is moved in the X and Y planes by means of two stepper motors. This arrangement is shown in Figure 100.

A simple algorithm for unidirectional scanning of a test piece could involve the following basic steps:

1. Moving the head to its datum position (coordinates 0,0).
2. Scanning a line at a time and, at each step in the X plane, reading and storing the data returned from the photodetector.
3. Returning the scanning head to its zero X axis coordinate at the end of each line and advancing the Y coordinate by one before scanning the next line.

An improved algorithm (using bidirectional scanning) could involve scanning alternate lines in different directions, advancing the Y co-ordinate by one as the end of each line is reached. This technique avoids the need to return the scanning head at the end of each line which is unproductive in terms of reading data from the test piece during the return period.

Displays and operator inputs

The following operator inputs will be required:

START Normally-open push button switch. Operated when the user has positioned the test piece and is ready to activate the scan. (The START button is to be inactive if the system is not ready for use.)

RESET Normally-open push button. Operated to abort a scan and return the sensing head to the datum position.

In order to provide the operator with an indication of status of the system the following displays will be required:

READY   Green LED to indicate that the system is ready for use.
DATUM   Red LED to indicate that the photohead has reached the datum position.
DETECT   Amber LED to indicate that the test piece has been detected by the photohead.

SCAN COMPLETE   Red LED to indicate that a scan is complete and that the data has been successfully stored on disk.
ERROR   Red LED to indicate that an error has occurred (for example disk full).

Configuration options

In order to simplify the process of making changes to the way the system behaves, we shall include a DIL switch which will be sensed during initialization. This switch which will determine the scale of the profile by determining the size of the horizontal and vertical steps. The switch will also be used to determine whether or not data will be stored in a disk file and also whether the scan produced will be unidirectional or bi-directional.

Hardware

In order to keep things as simple as possible, we shall use a standard 8-bit processor card based on a Z80 CPU. The card has 64 K RAM, a disk interface and two RS-232C serial ports. One of the RS-232C ports will be connected to a conventional VDU terminal while the other will be used to implement the serial data link to the remote computer installation. We

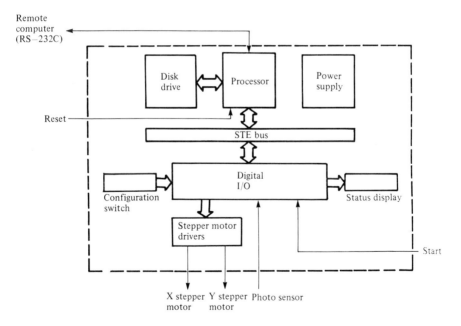

Figure 101 *Hardware configuration employed in the profile digitizer*

shall also require a digital I/O board and some specialized signal conditioning (along the lines described in Chapter 8) for driving the stepper motors which move the sensing head. The hardware will be mounted in a standard IEC rack enclosure fitted with a switched-mode power supply, single 3.5 in disk drive, and five-slot backplane. The hardware configuration is shown in Figure 101.

Software

This application requires a number of routines which can only be effectively written using assembly language. The application also calls for disk filing and the generation of prompts and messages which involve routines which can be more easily written in a high level language such as BASIC. As a compromise between these two conflicting requirements we shall use the C language. This provides the necessary file handling functions as well as allowing us to include in-line assembly language routines.

Program and data storage

The operating system and control software is expected to occupy no more than approximately 34 kilobytes of RAM. A single double-sided double-density 3.5 in disk (with a capacity of approximately 800 kilobytes) will be more than adequate for program and data storage. Since the data for a complete scan (1024 × 1024 sampling points) would normally occupy 128 kilobytes, a system of compression (described briefly later) can be usefully employed in order to minimize the size of disk files.

The operating system, control program, and various utilities will be provided on one (boot) disk. The disk will then be removed in order to permit the drive to be used for data storage during normal program execution. The user must, therefore, be prompted to remove the program disk at an appropriate point during program initialization.

I/O facilities

Since we are dealing with straightforward TTL compatible inputs and outputs, we shall make use of a single proprietary 40-channel digital I/O card of the type described in Chapter 3. I/O hardware will be attached to the ports provided on this card on the following basis:

DIP configuration switch.   Port C of PPII (configured for input). All 8 bits will be connected to switches but initially only bits 0, 1 and 2 will be used. Bit 0 will be used to determine whether a bidirectional scan will be performed. Bit 1 will be used to set the scale; when bit 1 returns a logic 0 the step size will be 0.5 mm otherwise the ×1 scale will be selected. Bit 2 will determine whether a disk data file will be produced. This facility can be used to inhibit disk output during testing. The DIL switch will thus operate according to Table 36.

Table 36

Bit 7 6 5 4 3 2 1 0	Disk file	Scale	Bidirectional scan
× × × × 0 0 0	No	×1	No
× × × × 0 0 1	No	×1	Yes
× × × × 0 1 0	No	×2	No
× × × × 0 1 1	No	×2	Yes
× × × × 1 0 0	Yes	×1	No
× × × × 1 0 1	Yes	×1	Yes
× × × × 1 1 0	Yes	×2	No
× × × × 1 1 1	Yes	×2	Yes

× = don't care.

START button.   Port A of PPI2 (configured for input). Bit 0 only will be used. A logic 0 will be returned when the START button is depressed.

RESET button.   Wired directly to the processor board using pins supplied available in the link area. The processor's $\overline{\text{RESET}}$ line will be activated when the button is depressed.

Status LEDs.   Port A of PPI1 (configured for output). Bits 0 to 4 will be used on the following basis:

LED	Bit
READY	0
DATUM	1
DETECT	2
SCAN COMPLETE	3
ERROR	4

Each line will be active when high (i.e. an LED will become illuminated when its respective port output line is taken high).

X stepper motor drive.   Port B of PPI1 (configured for output). Bits 0 to 2 will be used on the following basis:

Function	Bit
RESET	0
DIRECTION	1
STEP	2

Y stepper motor drive. Port C of PPI2 (configured for output). Bits 0 to 2 will be used on the following basis:

Function	Bit
RESET	0
DIRECTION	1
STEP	2

Photosensor. Port B of PPI2 (configured for input). Bit 0 only will be used.

I/O addresses

Using the default link settings of the 40-channel I/O board, the I/O port addresses in Table 37 result.

Table 37

Address	Register selected
E0	Port A of PPI1
E1	Port B of PPI1
E2	Port C of PPI1
E3	Control register for PPI1
E4	Port A of PPI2
E5	Port B of PPI2
E6	Port C of PPI2
E7	Control register for PPI1

Hence the various items of I/O hardware will be addressed as follows:

DIP configuration switch E2H (bits 0, 1 and 2).
START button   E4H (bit 0 only).
Status LEDs   E0H (bits 0 to 4).
X stepper motor drive   E1H (bits 0 to 2).
Y stepper motor driver   E6H (bits 0 to 2).

Initializing the digital I/O ports

The two 8255 PPI devices used in the digital I/O card will require initialization before use. It will be necessary to send a control word to each device and also send a byte to configure the buffers fitted to the board. The control words and buffer configuration bytes are selected from Table 38.

Table 38

| PPI1 direction | | | Control word (sent to PPI1's control register at address E3H) | Buffer configuration byte (sent to Port C of PPI2 at address E6H) |
Port A	Port B	Port C		
OUT	OUT	OUT	80H	YY11YYBA
OUT	OUT	IN	89H	YY11YYBA
OUT	IN	OUT	82H	YY01YYBA
OUT	IN	IN	8BH	YY01YYBA
IN	OUT	OUT	90H	YY10YYBA
IN	OUT	IN	99H	YY10YYBA
IN	IN	OUT	92H	YY00YYBA
IN	IN	IN	9BH	YY00YYBA

| PPI2 direction | | | Control word (sent to PPI2's control register at address E7H) | Buffer configuration byte (sent to Port C of PPI2 at address E6H) |
Port A	Port B	Port C		
OUT	OUT	OUT	80H	00XXBAXX
OUT	IN	OUT	82H	10XXBAXX
IN	OUT	OUT	90H	01XXBAXX
IN	IN	OUT	92H	11XXBAXX

Notes:
1 A is the enable bit for port A's buffer (1 to enable).
2 B is the enable bit for port B's buffer (1 to enable).
3 X refers to bits which control PPI1.

In this application we shall need to send control words of 89H and 92H to the control registers of PPI1 and PPI2 at respective addresses E3H and E7H. The configuration byte (found by combining the data from the two parts of Table 38) evaluates to FFH and this must be sent to port C of PPI2 at address E6H. A C function for initializing the I/O ports will thus take the form:

```
initport()
{
#asm
        MVI  A,89H
        OUT  0E3H
        MVI  A,92H
        OUT  0E7H
        MVI  A,0FFH
        OUT  0E6H
#endasm
}
```

Reading the configuration switch

Having initialized the two PPI devices we will probably wish to read the DIL configuration switch. This is achieved by the following function:

```
int s;

getconfig()
{
#asm
         IN  0E2H
         STA s
#endasm
}
```

The state of the configuration switch will be contained in the integer variable, s, and this will subsequently be used to determine the operational parameters of the system.

Bit 0 of the value in s determines whether or not directional scan is selected. The state of bit 0 (whether set or reset) can be tested using the if...else construct. The following code shows how:

```
int s;

if(s&1)
   {
      biscan();
   }
else
   {
      uniscan();
   }
```

If bit 0 is set, the condition following if evaluates to 1 (i.e., true) and the biscan( ) function is executed. If bit 0 is reset, the condition following if evaluates to 0 (i.e., false) and the program executes the uniscan( ) function.

Bit 1 of the value in s determines the step size. The value of bit 1 can similarly be tested using if...else:

```
int s;
int scale;

if(s&2)
   {
      scale = 2;
   }
else
   {
      scale = 1;
   }
```

If bit 1 is set, the condition following if evaluates to 1 (i.e., true) and the value of the integer scale is set to 2 otherwise it is set to 1. Bit 2 of the value in s determines whether or not a disk file is generated. The test is applied as follows:

```
int s;
if(s&4)
   {
      diskfile();
   }
```

Note that, since no alternative option is required, we do not need a matching else statement.

Stepper motor routines

The stepper motor lines are connected to the I/O port lines as follows:

Table 39

Motor	I/O address	Bit	Function
X	E1H	0	RESET
X	E1H	1	DIRECTION
X	E1H	2	STEP
Y	E6H	0	RESET
Y	E6H	1	DIRECTION
Y	E6H	2	STEP

The following C function will reset both stepper motor drivers:

```
int s;
resetmot()
{
#asm
        MVI  A,01H
        OUT  0E1H
        OUT  0E6H
#endasm
}
```

To produce single clockwise and anticlockwise steps on the X and Y motors the following functions would be used:

```
xcstep()
{
#asm
        MVI  A,04H
        OUT  0E1H
        MVI  A,00H
        OUT  0E1H
#endasm
}


xastep()
{
#asm
        MVI  A,06H
        OUT  0E1H
        MVI  A,02H
        OUT  0E1H
#endasm
}


ycstep()
{
#asm
        MVI  A,04H
        OUT  0E6H
        MVI  A,00H
        OUT  0E6H
#endasm
}


yastep()
{
#asm
        MVI  A,04H
        OUT  0E6H
        MVI  A,00H
        OUT  0E6H
#endasm
}
```

Note that each of the foregoing routines generates a transition on the STEP line while maintaining the desired logical states on the DIRECTION and RESET lines.

The size of a single step ultimately determines the resolution of the system. Since we need to provide two scales, we can generate two steps (transitions) when the ×2 scale is selected and use a single step (transition) when the system is operated in ×1 mode. The following C function shows how this can be achieved:

```
xmove()
{
   if (scale==1)
      {
         xcstep();
         delay(20);
      }
```

```
    else
        {
            xcstep();
            delay(20);
            xcstep();
            delay(20);
        }
}
```

Note that a short delay has been provided between steps in order that the stepper motor has time to move to the next position before a further step is initiated. The delay function provided in Chapter 7 can easily be modified for this purpose.

A function to scan a single line would take the form:

```
int xpos;

scanline()
{
   for (xpos=0;xpos<1024;++xpos)
    {
      xmove();
      detect();
    }
}
```

While a routine to perform a complete unidirectional scan would take the form:

```
int ypos;

uniscan()
{
   for (ypos=0;ypos<1024;++ypos)
    {
      scanline();
      store();
      xback();
      ymove();
    }
    yback();
}
```

Routines xback( ) and yback( ) respectively reset the position of the photosensor to the start of a line (xpos=0) and frame (ypos=0). The detect( ) function reads data returned from the photosensor and also records any changes in RAM, storing the current position of the photosensor.

Recording only changes in level (rather than the level detected at each coordinate) is instrumental in making a vast reduction in the size of disk files produced (an uncompressed digitized profile will require a disk file of 128 kilobytes whereas a typical compressed file requires less than 20 kilobytes).

The store( ) function appends the value of ypos to the data stored in RAM whenever no change is detected in the previous line. The function also maintains a pointer to the size of the data stored in RAM and periodically appends data to a disk file before resetting the pointer. This procedure is necessary in order to ensure that the data can be accommodated within the limited RAM space. The alternative (i.e., that of storing data for each scanned line at a time) is unattractive since the large number of disk accesses required will be instrumental in dramatically slowing the process down. It should be noted that the store( ) function is placed immediately before the xback( ) function so that a later enhancement (in the form of an intelligent stepper motor controller) will allow the disk access to take place during the period in which the photosensor is returned to its starting coordinate. This advantage will, of course, be lost in the case of a bidirectional scan which, in any event, should produce a faster scan.

An industrial weighing station

The need to accurately weigh items produced on a continuous basis is a common requirement of process control systems. This application shows how the STEbus can be used to control part of a continuous process permitting both fully-automatic and semi-automatic operation.

Specification

The specification calls for an automatic weighing station which can be

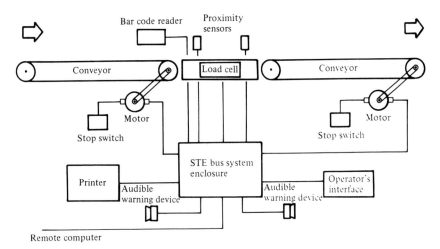

Figure 102 *Basic arrangement of the weighing station*

incorporated into a conveyor system along the lines shown in Figure 102. The station should provide an accuracy of better than ±1% of full-range (representing a maximum load of 100 kg). The weighing station should determine the weight of each item and transmit this, together with a code number (entered manually or with the aid of a bar code reader), to a remote supervisory computer installation. The distance to the remote computer is approximately 20 m and thus a conventional RS-232C link will be adequate. The weighing station should also be capable of producing hard-copy output to a conventional dot-matrix printer fitted with a Centronics parallel interface.

Inputs

The following six inputs to the system will be required:

1. Operator's keypad.
2. Bar code reader (via RS-232C).
3. Emergency stop switches (two to be fitted, one in the proximity of each conveyor).
4. Proximity sensors (fitted at the end of each conveyor).
5. Load cell.
6. System RESET button.

The operator's keypad will require a maximum of sixteen keys. The ten number keys will be used to input the item code in the event that the automatic bar code reader fails to detect the bar code. The remaining six keys will be used as follows:

AUTO	Pressed to select automatic operation.
START	Pressed to start the conveyor system.
STOP	Pressed to stop both conveyors.
ENTER	Pressed to terminate manual code entry.
MANUAL	Pressed to select manual code entry.
DISPLAY	Removes current prompt from the display and shows code number of last item together with its weight. Also sends this data to a printer (if connected).

The two emergency STOP buttons are to be capable of interrupting the supplies to the conveyor motors independently of the bus-based controller. Each must be fitted in series with the AC power line to the motor in question.

Outputs

The following outputs will be required:

1. LCD message display panel (40 columns × 2 lines).

2 Audible warning devices (fitted at the end of each conveyor).
3 Conveyor motor control (via two solid state relays).
4 Parallel (Centronics) printer.

Hardware

Since the application is not particularly demanding in terms of processor power, a simple Z80-based processor card will be used. This card has sockets for up to 32 kilobytes of ROM and 16 kilobytes of static RAM and also two RS-232C serial ports (one of which will provide the serial data link to the remote computer). A digital I/O board will be required, as will an analogue input board offering 12-bit resolution. Specialized input signal conditioning will be required to interface the load cell to this latter board. The solid-state relays and audible warning devices will require the services of a specialized output driver (described in detail later).

A watchdog controller will be incorporated to perform a variety of functions including a means of connecting the RESET switch, an interface for the LCD display, and a Centronics compatible printer port. The system hardware will be mounted in a standard IEC rack enclosure fitted with a switch-mode power supply and five-slot backplane. The LCD display and keypad will be mounted in a separate sloping front enclosure attached to the system unit by means of a multiway flexible cable. The system hardware configuration is depicted in Figure 103.

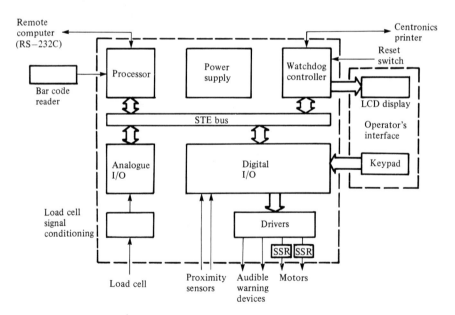

Figure 103 *Hardware configuration employed in the weighing station*

Software

Since no part of the process is time critical and since the I/O routines are relatively straightforward, the program can will be developed in BASIC and subsequently compiled for transfer to ROM in order to permit stand-alone operation.

Program and data storage

Program development will take place on a simple disk-based STE development system comprising a Z80 processor, power supply, dual disk drive, CP/M and B80 BASIC compiler. After testing and debugging, the program can be compiled and transferred to ROM using an STE EPROM programmer. The compiled BASIC program will require approximately 7 kilobytes of memory and can thus be accommodated within a single 2764 (8 kilobytes) EPROM. The procedure for transferring the program to EPROM is fairly complex but can be broken down into the following steps:

1. Load the B80 compiler from disk.
2. Load the debugged BASIC program using a command of the form:

    load 'prog.bas'

3. Compile the program and generate an Intel format hex. file using a command of the form:

    comp 'prog', &400,&4000,$D000

    The parameters following the program name (prog) respectively indicate the address of the start of the run-time system, the address of the start of the program, and the highest address in usable RAM.
4. Exit to the system and invoke the CP/M debug utility, SID.
5. Load the hexadecimal file produced in step 3 into memory using the SID command:

    rprog.hex

    After a successful load, SID will display the end address of the hexadecimal file under the heading 'NEXT'. Assuming that this address appears as 5D00, a binary file should now be created using a command of the form:

    wprog.epr,4000,5D00

6. The user should now return to the operating system (using CTRL-C) and invoke the EPROM utility program using a command of the form:

    eprom

    The user will be prompted for the type of EPROM and the disk filename to be used. The response to this second query should take the form:

prog.epr

The user is then further prompted for data start and end addresses (the values provided by the 'Data loaded' message should be used) as well as a ROM start address (this will usually be 0).

The user is finally prompted for the programming algorithm to be employed (fast or slow) and the required programming voltage (typically 21 V).

7 If the EPROM has been successfully programmed an appropriate message will be generated to indicate that no errors have occurred during programming. The EPROM may then be removed from the EPROM programmer and placed in the target processor board. Note that the B80 run-time package (supplied in ROM) must also be fitted to the target board.

I/O configuration

The various I/O devices will be connected as follows:

Operator's keypad.   Connected to Port A of PPI2 (configured for input). Bits 0 to 3 will be used for the key code while bit 7 will be connected to the data available (DA) line. When bit 7 is set, this will indicate that a keypress is available.

Proximity sensors.   Connected to Port C of PPI1 (configured for input). Bits 0 and 1 only will be used.

Audible warning device.   Connected to Port A of PPI1 (configured for output). Bits 0 and 1 will be used.

Bar code reader.   Connected to the second RS-232C port (channel B) on the processor card.

Load cell.   Connected to the analogue I/O board.

LCD display.   Connected to the LCD port available on the watchdog controller.

Printer.   Connected to the Centronics port available on the watchdog controller.

Solid state relays.   Connected to Port A of PPI1 (configured for output). Bits 2 and 3 will be used.

I/O addresses

Using default digital I/O port addresses and configuring the analogue I/O card to addresses 7EH and 7FH gives the following I/O assignment:

Load cell   7EH (control and status).
                      7FH (data).

Printer	BCH (data).
	BDH (status and control).
LCD display	BEH (control and status).
	BFH (data).
Audible warning devices	E0H (bits 0 and 1).
Solid-state relays	E0H (bits 2 and 3).
Proximity sensors	E2H (bits 0 and 1).
Keypad	E4H (bits 0 to 3 and bit 7).

Initializing the digital I/O ports

Since the direction of data transfer to and from the I/O ports is the same as that used in the previous example, a similar initialization routine will be required. This can be achieved directly from BASIC without the need to revert to assembly language modules. The following routine shows how:

```
10 REM Initialise digital I/O ports
15 OUT (&E3),&89 : REM PPI1 A and B output, C input
20 OUT (&E7),&92 : REM PPI2 A and B input, C output
25 OUT (&E6),&FF : REM Enable buffers
```

The LCD display also requires initialization. This is achieved by sending a sequence of control bytes to address BEH (190 decimal). The byte values (given in decimal) shown in Table 40 are required.

Table 40

Byte (decimal)	Function
56*	Set 8-bit wide interface mode (sent several times to recover from power interruption).
15	Display on, cursor flashing.
6	Cursor direction, memory increment.
128	Start at beginning of display RAM.
1	Clear display and home cursor.

*Note that this value should be 48 for a one-line display.

The required LCD initialization can be performed using the following routine:

```
30 REM Initialise LCD
35 RESTORE 19000
40 FOR i%=1 TO 7
45 READ z%
50 OUT 190,z%
55 NEXT i%

19000 DATA 56,56,56,15,6,128,1
```

Reading the keypad

The keypad can be easily read using the multitasking feature of BASIC-Z80. This feature allows two tasks to run concurrently. The alternate task is defined and the program is instructed to swap tasks on the start of each line of source code, by incorporating the following statements within the initialization routine:

```
60 TASK poll_keys
65 ON SCAN SWAP TASK
```

The alternate task effectively polls the keyboard and detects whether a key is currently being depressed. The procedure takes the form:

```
8000 DEF poll_keys
8010 x%=INP(&E4)
8020 IF x%<127 THEN ENDPROC
8030 IF x%=.......
................
```

If no key is currently depressed the procedure terminates in line 8020. The lines from 8030 onwards detect a specific key depression and take appropriate action.

Displaying prompts and messages

Prompts and messages can be very easily generated on the LCD display and require relatively few lines of BASIC. The messages required can all be defined within the same block of code:

```
80 REM Prompts and messages
81 version$="*** WEIGHING STATION V1.1 ***"
82 auto$="Automatic operation selected"
83 manual$="Manual operation selected"
84 fail$="Bar code not valid - please enter"
85 invalid$="Code not valid - please try again"
86 cancel$="Press <MANUAL> to cancel"
87 start$="Press <START> to continue"
88 code$="Code for this item: "
89 weight$="Weight for this item: "
```

Messages can later be sent to the display by passing the required string into a message handling procedure. The method is illustrated below:

```
7000 REM Display message
7010 DEF message(display$)
7020 LOCAL i%, display$
7025 OUT 190,1 : REM clear display
7030 FOR i%=0 TO LEN(display$)
7040 OUT 191,ASC(MID$(display$,i%,1))
7050 NEXT i%
7060 ENDPROC
```

To print the title and version number of the program during initialization, a line of the form:

```
60 message(version$)
```

can be used. The displayed message can be easily changed using such procedure calls as:

```
345 message(start$)
```

String variables can also be combined with predefined message strings prior to making a call to the message display procedure. The following code, where c$ is a string formed from the data returned by the bar code reader, shows how:

```
440 .........
450 cd$=code$+c$
460 message(cd$)
470 .........
```

Driving the solid-state relays and audible warning devices

The solid-state relays and audible warning devices use the same output port. To provide additional output current drive, a UCN 5801 octal Darlington output driver is used, as shown in Figure 104.

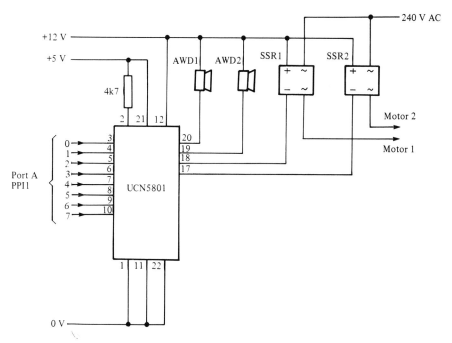

Figure 104 *Output driver arrangement used in the weighing station*

Device controlled:	SSR2	SSR1	AWD2	AWD1
Bit number:	3	2	1	0

As an example, suppose we wish to simultaneously operate AWD1 and AWD2. The corresponding output bit pattern would be 00000011 (equivalent to a decimal value of 3). This output could be produced by a line of the form:

```
2110 OUT &E0,3
```

Alternatively, to operate SSR1 and AWD2 we would need to output a bit pattern of 00000101 (decimal 5). This output could be produced by a line of the form:

```
2125 OUT &E0,5
```

Driving the printer

The Centronics port provided on the watchdog controller may be easily controlled from within BASIC. When a status byte is written to port address BCH (188 decimal) the Centronics STROBE line is pulsed low. A Centronics ACKNOWLEDGE signal causes bit 4 of the status byte from port address BDH (189 decimal) to be pulsed low. Bit 3 of the status byte is used to convey the Centronics BUSY signal. Bits 2 and 1 of the status byte are used to indicate PAPER END and printer OFF-LINE respectively. The status byte should be read by software before data is sent to the printer at address BCH. The following code shows how:

```
7770 st=INP(189)
7775 IF (st AND 8)=8 THEN message(" Printer busy! ")
7780 IF (st AND 4)=4 THEN message(" Printer out of paper! ")
7785 IF (st AND 2)=2 THEN message(" Printer is off-line! ")
```

Finally, the following code shows how data (in the form of string pr$) can be sent to the printer:

```
7820 FOR i%=1 TO LEN(pr$)
7825 st=INP(189) AND 8: REM get status and
7830 IF st=8 GOTO 7825: REM loop until ready
7835 byte$=MID$(pr$,i%,1)
7840 OUT 188,ASC(byte$)
7845 NEXT i%
```

10
Reliability and fault-finding

Ask designers of process control systems to define their objectives and their reply will probably involve optimizing system performance within the constraints imposed by time and a given budget. At the same time, they will not wish to compromise the overall quality or reliability of the system. This final chapter deals with quality and reliability in the context of bus-based process control systems and will also examine some basic fault-finding techniques applicable to bus-based systems which can be instrumental in reducing system down-time.

Quality procedures

In a general engineering context, quality is often defined as the degree to which a product or its components conform to the standards specified by the designer. Such standards generally relate to identifiable characteristics relating to materials, dimensions, tolerances, performance and reliability. In a production engineering environment, the degree of effectiveness in meeting these standards can be assessed by conventional acceptance tests, sampling and statistical analysis. Readers will undoubtedly be familiar with one or more of these techniques. In the case of a one-off process control systems, quality control procedures will generally involve the following tests:

1. Functional tests under normal (or simulated normal) operating conditions.
2. Functional tests under extreme (or simulated extreme) operating conditions.
3. Overload tests to determine the behaviour of the system under abnormal or totally unexpected operating conditions.
4. Environmental testing to determine the performance of the system

under various extreme conditions of humidity, temperature, vibration, etc.

Process control specialists must inevitably undertake some or all of the functions performed by the quality engineer in a production environment. Not only will they be involved with specifying, designing, building and installing a system but they must also ensure that the overall quality of the system is assured and that the system meets the standard and criteria laid down in the initial specification. The quality assurance function requires an ongoing involvement with the project from design to subsequent installation and use.

Reliability and fault-tolerance

Reliability of a process control system is often expressed in terms of its percentage *up-time*. Thus, a system which is operational for a total of 950 hours in a period of 1000 hours is said to have a 95 per cent up-time. An alternative method of expressing reliability involves quoting a 'mean time before failure' (MTBF). The MTBF is equivalent to the estimated number of hours that a system is expected to operate before it encounters a failure requiring a period of *down time*.

Various techniques can be used to make bus-based process control systems inherently fault tolerant. Such techniques can be classified under the general categories of *hardware* and *software*. We shall discuss these techniques separately.

Hardware techniques

Hardware methods generally involve the use of a *watchdog controller*. This device is a standard bus card which contains various hardware devices for monitoring the performance of the system. The facilities available will generally include:

- Generating a status byte which is periodically read (typically every 1 to 2.5 seconds) by the software in order to ascertain the state of the system. If the status byte is not read within a predetermined period, the watchdog controller assumes that a fault condition has been encountered and the board takes appropriate action (for example, generating a system reset by taking the SYSRST* line low). Typical conditions which result in the watchdog status byte not being read involve attempts to access missing hardware (a *hardware hang*) or the software executing an infinite loop (a *software hang*).
- Monitoring one or more of the bus power rails and generating appropriate signals when the voltage present fails to meet the defined

tolerance limits for the rail concerned. Typical actions involve asserting the ATNRQ0* and SYSRST* lines or activating external relays to permit supply changeover.
- The ability to accept an external signal (via the contacts of a push-button switch, relay or isolator) which will generate a system reset by taking SYSRST* low.
- Fitting an uninterruptible power supply.
- Exploiting the multiprocessor capability of the STE bus by making use of independent processors and, where necessary, duplicate I/O circuitry attached to independent signal conditioning boards.

Bus timeout

Most users of the STE bus will sooner or later encounter an unfortunate feature of asynchronous bus systems in the shape of a hardware hang. Since these can have dire consequences, it is worth explaining how they arise and how a watchdog device can be used to counteract them.

The fact that the STE bus is asynchronous means that a bus master (i.e. CPU) will start a bus transfer by asserting DATSTB*) and wait for an acknowledge signal (DATACK*) from the addressed slave board. However, if there is no peripheral board configured to the I/O address generated by the master, the CPU will 'hang'; waiting indefinitely for an acknowledge signal. To prevent this, bus timeout circuitry within the watchdog controller monitors the bus and checks whether a corresponding DATACK* acknowledge signal has been generated in response to the DATSTB* signal produced by the master. If the DATACK* signal does not appear within a predetermined period, the watchdog device asserts a transfer error (TRFERR*) signal in order to release the CPU from its hung state. Typical waiting times can be set to between 8 μs and 16 μs (longer times are applicable in the case of slow peripheral devices such as floppy disk controllers).

Software techniques

Software techniques generally involve incorporating software routines, procedures, or functions which will:
- Perform full system diagnostic tests during initialization.
- Perform periodic diagnostic tests during program execution (for example, periodically reading a status byte).
- Ensure that out of range indications are recognized and erroneous data is ignored.
- Generate error and warning messages to alert the user to the presence of a malfunction.

- Log faults as they occur together, where possible, with sufficient information (including date and time) so that the user can determine the point at which the fault occurred and the circumstances prevailing at the time.

System software invariably incorporates simple diagnostic routines of the type mentioned above. The routines provided usually check both ROM and RAM devices. In the former case, the contents of each byte of ROM are successively read and a checksum is generated. This checksum is then compared with a stored checksum. Alternatively, bytes are incorporated within the ROM in order that the checksum generated equates to zero. If any difference is detected, initialization is aborted and an appropriate error message is generated.

In the case of RAM diagnostics, a different technique is employed. Here the process involves writing and reading each byte of RAM in turn. Checking, in each case, that the desired change has been effected. Where a particular bit refuses to be changed, the diagnostic procedure is temporarily halted and an appropriate error message is generated (this may also provide sufficient information for it to be possible to locate the individual device which has failed).

RAM diagnostics can also be carried out on a non-destructive basis. In such cases, the byte read from RAM is replaced immediately after each byte has been tested. It is thus possible to perform a diagnostic routine some time after the system has been initialized.

Fault-finding

Process control specialists will rarely wish to deal with fault-finding down to component level. In order to avoid a prohibitive investment in test equipment and technical expertise, it is generally considered more cost-effective to have such repairs carried out by specialists. Despite this, it is sometimes essential to minimize the time taken to correct the failure of a bus-based process control system. An ability to make on-site repairs, at least to board level, is thus highly desirable.

At first sight, the prospect of fault-finding a bus system can be somewhat daunting. This is especially true when those having to carry out the repairs may be relatively unfamiliar with electronic circuitry. However, in the author's experience, the vast majority of faults are attributable to failure of external devices (such as sensors, cables and connectors) rather than with the boards and cards themselves. Furthermore, even when dealing with boards within the system enclosure, most faults can be detected without recourse to sophisticated test gear.

When component rather than board level servicing has to be undertaken, it is useful to obtain a circuit diagram and service information on the equipment before starting work. This information will be invaluable when identifying components and establishing their function within the system as a whole.

Certain 'stock faults' (such as chip failure) may be prevalent on some boards and these should be known to manufacturers and their service agents. A telephone enquiry, describing the symptoms and clearly stating the type and version number of the card or board, will often save much time and effort. Furthermore, manufacturers are usually very receptive to information which leads to improvement of their products and may also be prepared to offer retrofit components and/or circuit modifications to overcome commonly identified problems.

Test equipment

A few items of basic test gear will be required by anyone attempting to perform fault location on bus systems. None of the basic items is particularly costly and most will already be available in an electronics laboratory or workshop. For the benefit of the newcomer to electronics we will briefly describe each item and explain how it is used in the context of bus system fault-finding.

Multirange meters

Multirange meters provide either analogue or digital indications of voltage, current, and resistance. Such instruments are usually battery powered and are thus eminently portable. Connection to the circuit under test is made via a pair of test leads fitted with probes or clips. The following specification is typical of a modern digital multirange meter:

DC voltage	200 mV, 2 V, 20 V, 200 V, and 1.5 kV full-scale accuracy ± 0.5%. Input resistance 10 MΩ.
AC voltage	2 V, 20 V, 200 V and 1 kV full-scale accuracy ± 2%. Input resistance 10 MΩ.
DC current	200 µA, 2 mA, 20 mA, 200 mA and 2 A full-scale accuracy ± 1%.
AC current	200 µA, 2 mA, 20 mA, 200 mA and 2 A full-scale accuracy ± 2%.
Resistance	200 Ω, 2 kΩ, 20 kΩ, 200 kΩ, and 20 MΩ full-scale accuracy ± 2%.

A typical application for a multirange meter is that of checking the various supply voltages present on the bus. For an operational STE bus system the supply voltages should be within the range given in Table 41.

Table 41

Supply	Nominal value	Acceptable value minimum	maximum	Connector pin number
$+V_{cc}$	+5 V	+4.87 V	+5.25 V	31A and 31C
+AUX V	+12 V	+11.4 V	+12.6 V	30C
−AUX V	−12 V	+11.4 V	−12.6 V	30A
+VSTBY	+5 V	+4.75 V	+5.25 V	29C

Multirange meters may also be used for checking the voltages present on the supply rails within individual bus cards. Particular points of interest will be those associated with the supplies to individual chips. In such cases, extender boards can be employed in order to gain access to the card while it is still connected to the backplane. Extenders may also incorporate labelled test points which facilitate connection to the bus lines and thus avoid the need for rear access to the backplane.

Multirange meters may even be used to display logic states on signal lines which remain static for long periods. This is often the case when dealing with I/O lines however, in situations where logic levels are continuously changing, a multimeter cannot provide a reliable indication of the state of a line.

Where logic levels do remain static for several seconds, the multirange meter may be used on the DC voltage ranges to sense the presence of logic 0 or logic 1 states according to Table 42 which gives the conventional voltage levels associated with TTL logic.

Table 42

Logic level	Voltage present
1	>2.0 V
0	<0.8 V
Indeterminate	0.8 V to 2.0 V

It should be noted that an 'indeterminate' logic level may result from a tristate condition in which bus drivers are simultaneously in a high impedance state. Modern high impedance instruments will usually produce a misleading fluctuating indication in such circumstances and this can sometimes be confused with an actively pulsing bus line.

Logic probes

The simplest and most convenient method of tracing logic states involves the use of a logic probe rather than a multirange meter. This invaluable tool comprises a hand-held probe fitted with LEDs to indicate the logical state of its probe tip.

Unlike multirange meters, logic probes can generally distinguish between lines which are actively pulsing and those which are in a permanently tristate condition. In the case of a line which is being pulsed, the logic 0 and logic 1 indicators will both be illuminated (though not necessarily with the same brightness) whereas, in the case of a tristate line neither indicator should be illuminated.

Logic probes generally also provide a means of displaying pulses having a very short duration which may otherwise go undetected. A pulse stretching circuit is usually incorporated within the probe circuitry so that an input pulse of very short duration is elongated sufficiently to produce a visible indication on a separate pulse LED.

Logic probes invariably derive their power supply from the circuit under test and are connected by means of a short length of twin flex fitted with insulated crocodile clips. While almost any convenient connecting point may be used, the leads of an electrolytic +5 V rail decoupling capacitor or the +5 V and GND terminals of the bus backplane both make ideal connecting points which can be easily identified.

Logic pulsers

It is sometimes necessary to simulate the logic levels generated by a peripheral device or sensor. A permanent logic level can easily be generated by pulling a line up to +5 V via a 1 kΩ resistor or by temporarily tying a line to 0 V. However, on other occasions, it may be necessary to simulate a pulse rather than a permanent logic state and this can be achieved by means of a logic pulser.

A logic pulser provides a means of momentarily forcing a logic level transition into a circuit regardless of its current state and thus overcomes the need to disconnect or desoldering any of the devices. The polarity of the pulse (produced at the touch of a button) is adjusted so that the node under investigation is momentarily forced into the opposite logical state. During the period before the button is depressed and for the period after the pulse has been completed, the probe tip adopts a tristate (high impedance) condition. Hence the probe does not permanently affect the logical state of the point in question.

Pulsers derive their power supply from the circuit under test in the same manner as logic probes. Here again, the leads of an electrolytic decoupling capacitor or the +5 V and GND terminals fitted to the bus backplane make suitable connecting points.

Oscilloscopes

The use of an oscilloscope in the examination of time related signals (waveforms) will already be well known to the vast majority of readers. Such instruments provide an alternative means of tracing logic states present in a bus system and may also be used for detecting noise and unwanted AC signals which may be present on power supply rails. It must, however, be stressed that, since low cost oscilloscopes generally do not possess any means of storing incoming signals, severe triggering problems arise when signals are non-repetitive. This is an important point since many of the digital signals present on a bus are both asynchronous and non-repetitive.

Apart from displaying the shape of waveforms present in a bus system, oscilloscopes can also be used to make reasonably accurate measurements of voltage and time. In such cases, measurements are made with reference to a graticule fitted to the CRT and scale factors are applied using the time and voltage range switches. However, before attempting to take measurements from the graticule it is essential to check that any variable front panel controls are set to the calibrate (CAL) position. Failure to observe this simple precaution may result in readings which are at best misleading or at worst grossly inaccurate.

Since modern oscilloscopes employ DC coupling throughout the vertical amplifier stages, a shift along the vertical axis will occur whenever a direct voltage is present at the input. When investigating waveforms in a circuit one often encounters AC signals superimposed on DC levels; the latter may be removed by inserting a capacitor in series with the input using the AC-GND-DC switch. In the AC position the capacitor is inserted at the input, whereas in the DC position the capacitor is shorted. If GND is selected the vertical input is taken to common (0 V) and the input terminal is left floating. In order to measure the DC level of an input signal, the AC-GND-DC switch must first be placed in the GND position. The vertical position is then adjusted so that the trace is coincident with the central horizontal axis. The AC-GND-DC switch is then placed in the DC position and the shift along the vertical axis measured in order to ascertain the DC level.

Most dual-beam oscilloscopes incorporate a chopped-alternate switch which selects the mode of beam splitting. In the 'chopped' position, the trace displays a small portion of one vertical channel waveform followed by an equally small portion of the other. The traces are thus sampled at a fast rate so that the resulting display appears to consist of two apparently continuous traces. In the 'alternate' position, a complete horizontal sweep is devoted to each channel on an alternate basis.

Chopped mode operation is appropriate to signals of relatively low frequency (i.e., those well below the chopping rate) where it is important that the display accurately shows the true phase relationship between the two displayed signals. Alternate mode operation, on the other hand, is suitable for high frequency signals where the chopping signal would otherwise corrupt the display. In such cases it is important to note that the relative phase of the two signals will not be accurately displayed.

Most modern oscilloscopes allow the user to select one of several signals for use as the timebase trigger. These *trigger source* options generally include an internal signal derived from the vertical deflection system, a 50 Hz signal derived from the AC mains supply, and a signal which may be applied to an *external trigger input*. As an example, the 50 Hz trigger source should be selected when checking for mains borne noise and interference whereas the external trigger input may usefully be derived from a processor clock signal when investigating the synchronous signals present within the local bus used by a processor.

Fault location procedure

To simplify the process of fault location on a complex bus system, it is useful to consider the system as a number of interlinked subsystems. Each subsystem can be further divided into its constituent elements. Fortunately, the use of a bus makes fault-finding very straightforward since it is eminently possible to isolate a fault to a particular part of the system just by removing a suspect board and substituting one which is known to be functional.

In order to assist readers, the following eight point checklist has been provided. The questions should be answered *before* attempting to make any measurements or remove any suspect boards.

1. Has the system operated in similar circumstances without failure? Is the fault inherent in the system?
2. If an inherent fault is suspected, why was it not detected by normal quality procedures?
3. If the fault is not considered inherent and is attributed to component failure, in what circumstances did the equipment fail?
4. Is the fault intermittent or is it present at all times?
5. If the fault is intermittent, in what circumstances does it arise? Is it possible to predict when the fault will occur?
6. To facilitate testing and diagnosis, can conditions be reproduced so that the fault manifests itself permanently?
7. What parts of the equipment are known to be functioning correctly? Is it possible to isolate the fault to a particular part?
8. Is the fault a known 'stock fault'? Has the fault been documented elsewhere?

Having answered the foregoing questions, and assuming that one is confronted with a system which is totally unresponsive, the first step is that of checking the power supply rails using a multirange meter. Where any one of the supply rails is low (or missing altogether) the power supply should be disconnected from the backplane and the measurement should be repeated in order to establish whether the absence of power is due to failure of the power supply or whether the fault can be attributed to excessive loading. This, in turn, can either be due to a short-circuit component failure within a bus card or a similar fault in the backplane.

Most bus systems employ switched mode power supplies and it should be borne in mind that such units generally require that a nominal load be present on at least one of their output rails before satisfactory regulation can be achieved. Failure to observe this precaution can lead the unsuspecting test engineer to conclude that a unit is not regulating correctly when it has been disconnected from a backplane or when all of the cards have been removed from a frame. In any event, it is advisable to consult the manufacturer's data before making measurements on switched mode power supplies.

Having ascertained that the bus system is receiving its correct power supply voltages, the next stage is that of activating the system reset line (SYSRST*) and noting whether any changes are produced. If a disk drive is fitted to the system, this will normally become active at this time. Alternatively, some form of prompt or 'system ready' indication may be produced. If neither of these indications is produced, it is worth removing the processor board and replacing it with a known functional unit. If this is not possible, the processor card should be removed and fitted to an extender. The extender should then be inserted into the slot vacated by the processor card (Figure 105). Conditions on the following signal lines should then be examined in turn using a logic probe or oscilloscope:

- SYSRST* Check that this line is pulsed low when the reset switch is depressed.
- SYSCLK Check that the signal on this line is a 16 MHz square wave.
- DATSTB, DATACK* and ADRSTB* In normal operation these lines will be regularly pulsing.
- TRFERR* This line should be high in normal operation.
- CM0–CM2 When CM2 is high, the other command modifier lines act as follows:
  CM0 is taken low during write operations but is high during read operations.
  CM1 is taken low when I/O is referenced but is high when memory is addressed.
- ATNRQ0*–ATNRQ7* These lines are used for interboard signalling and activity will vary depending upon the particular configuration employed.

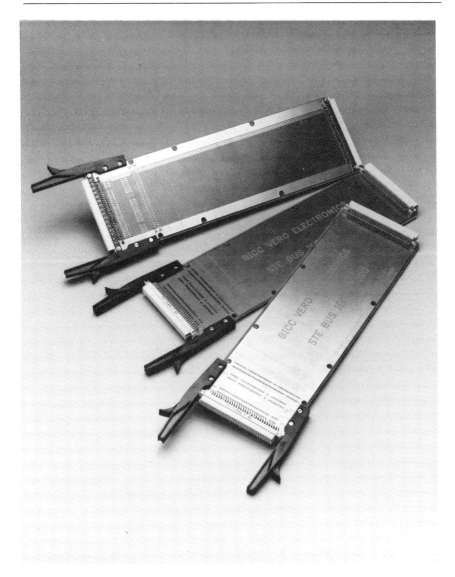

Figure 105 *Bus extenders (courtesy of BICC-Vero Electronics Ltd)*

- BUSRQ0*–BUSRQ1* and BUSAK0*–BUSAK1* These lines will only be active if an arbiter and several masters are present.

The state of the bus lines can often be instrumental in pointing to a faulty board. However, where a fault is intermittent (for example the system runs for a time before stopping) it is worth checking connectors and also investigating the cleanliness of the supply. Connectors are often prone to failure and, if the principal chips are socketed these, too, can cause problems. Intermittent faults can sometimes be corrected simply by pressing each of the larger chips into its socket. In some cases it may be necessary to carefully remove the chips before replacing them; the action of removal and replacement can sometimes be instrumental in wiping the contacts clean.

Where a fault is permanently present and one or more of the supply rail voltages is lower than normal, chip failure may be suspected. In such an event, the system should be left running for some time and the centre of each chip should be touched in turn in order to ascertain its working temperature. If a chip is running distinctly hot (i.e., very warm or too hot to comfortably touch) it should be considered a prime suspect. Where possible the temperature should be compared with that generated by a similar chip fitted in the same board or that present in another functional module. Where the larger chips have been fitted in sockets, each should be carefully removed and replaced in turn (disconnecting the power, of course, during the process) before replacing it with a known functional device.

Complications often arise when a bus system is configured for multiprocessing and several potential bus masters are present. If a processor is unable to gain access to the bus it may hang because another master already has control of the bus and has not released it. In such an event it will be necessary to check the BUSRQ* and BUSAK* lines using a logic probe or an oscilloscope. If the lines are enabled, check the control modifier lines to ensure that the bus cycle is not an acknowledge cycle and that the unsuccessful processor is asserting the data strobe line. If DATSTB* has been asserted, a DATACK* or TRFERR* signal should have been received from the slave card. If these signals have not appeared, check that the slave card responds to the address that the processor is generating.

Problems in multiprocessing systems also arise when more than one board is generating SYSCLK. Thus, where multiple masters are present, it is important to disable the SYSCLK facility on all but one board. Furthermore, slave I/O boards are invariably fitted with links which provide selection of addresses and attention request lines. These links should be adjusted so that no conflicts occur. Furthermore, when new or replacement cards are to be fitted to a system, it is essential to ensure that the links are correctly set before fitting.

Finally, in a perfect world there would be no uncertainty nor any ambiguity about the logic levels present in a digital system. Unfortunately, this is seldom the case since spurious signals (or 'noise') are invariably present to some degree. The ability to reject noise is thus an important requirement of bus-based control systems. This is particularly true where a system is to be used in a particularly noisy environment (such as a shipyard or steelworks). In such a situation, special precautions may be necessary in order to avoid corruption of signals and data and one or more of the following techniques may be aplied:

- Using a 'clean' AC supply for the bus system or, where such a supply is not available, fitting a supply filter or AC power conditioner.
- Screening all signal cables and returning the outer braid screen to earth (note that noise rejection is sometimes enhanced if the screen is only earthed at one point).
- Ensuring that the bus system enclosure is adequately earthed and that none of the outer panels or metal chassis parts are allowed to 'float'.
- Decoupling supply rails at the point at which they enter each signal conditioning board.
- Using adequately rated wiring between the power supply, backplane, and peripheral devices (for example, disk drives). The main +5 V and GND wiring, in particular, should make use of 24/0.2 mm (0.75 mm^2) or 32/0.2 mm (1.0 mm^2) stranded tinned copper conductors.
- In extreme cases, making use of optical fibres rather than twisted pairs or coaxial cables for transmission of digital signals.

Appendix 1
STE backplane signals

Pin	Row A	Row C
1	0V	0V
2	+5V	+5V
3	D0	D1
4	D2	D3
5	D4	D5
6	D6	D7
7	A0	0V
8	A2	A1
9	A4	A3
10	A6	A5
11	A8	A7
12	A10	A9
13	A12	A11
14	A14	A13
15	A16	A15
16	A18	A17
17	CM0	A19
18	CM2	CM1
19	ADRSTB*	0V
20	DATACK*	DATSTB*
21	TRFERR*	0V
22	ATNRQ0*	SYSRST*
23	ATNRQ2*	ATNRQ1*
24	ATNRQ4*	ATNRQ3*
25	ATNRQ6*	ATNRQ5*
26	0V	ATNRQ7*
27	BUSRQ0*	BUSRQ1*
28	BUSAK0*	BUSAK1*
29	SYSCLK	+VSTBY
30	−12V	+12V
31	+5V	+5V
32	0V	0V

Appendix 2
Decimal, binary, hexadecimal and ASCII conversion table

Decimal	Binary	Hexadecimal	ASCII
0	0	0	NUL
1	1	1	SOH
2	10	2	STX
3	11	3	ETX
4	100	4	EOT
5	101	5	ENQ
6	110	6	ACK
7	111	7	BEL
8	1000	8	BS
9	1001	9	HT
10	1010	A	LF
11	1011	B	VT
12	1100	C	FF
13	1101	D	CR
14	1110	E	SO
15	1111	F	SI
16	10000	10	DLE
17	10001	11	DC1
18	10010	12	DC2
19	10011	13	DC3
20	10100	14	DC4
21	10101	15	NAK
22	10110	16	SYN
23	10111	17	ETB
24	11000	18	CAN
25	11001	19	EM
26	11010	1A	SUB
27	11011	1B	ESC
28	11100	1C	FS
29	11101	1D	GS
30	11110	1E	RS
31	11111	1F	US
32	100000	20	
33	100001	21	!
34	100010	22	"
35	100011	23	£
36	100100	24	$
37	100101	25	%

38	100110	26	&
39	100111	27	'
40	101000	28	(
41	101001	29	)
42	101010	2A	*
43	101011	2B	+
44	101100	2C	,
45	101101	2D	−
46	101110	2E	.
47	101111	2F	/
48	110000	30	0
49	110001	31	1
50	110010	32	2
51	110011	33	3
52	110100	34	4
53	110101	35	5
54	110110	36	6
55	110111	37	7
56	111000	38	8
57	111001	39	9
58	111010	3A	:
59	111011	3B	;
60	111100	3C	<
61	111101	3D	=
62	111110	3E	>
63	111111	3F	?
64	1000000	40	@
65	1000001	41	A
66	1000010	42	B
67	1000011	43	C
68	1000100	44	D
69	1000101	45	E
70	1000110	46	F
71	1000111	47	G
72	1001000	48	H
73	1001001	49	I
74	1001010	4A	J
75	1001011	4B	K
76	1001100	4C	L
77	1001101	4D	M
78	1001110	4E	N
79	1001111	4F	O
80	1010000	50	P
81	1010001	51	Q
82	1010010	52	R
83	1010011	53	S
84	1010100	54	T
85	1010101	55	U
86	1010110	56	V
87	1010111	57	W

Dec	Binary	Hex	ASCII
88	1011000	58	X
89	1011001	59	Y
90	1011010	5A	Z
91	1011011	5B	[
92	1011100	5C	\
93	1011101	5D	]
94	1011110	5E	^
95	1011111	5F	_
96	1100000	60	`
97	1100001	61	a
98	1100010	62	b
99	1100011	63	c
100	1100100	64	d
101	1100101	65	e
102	1100110	66	f
103	1100111	67	g
104	1101000	68	h
105	1101001	69	i
106	1101010	6A	j
107	1101011	6B	k
108	1101100	6C	l
109	1101101	6D	m
110	1101110	6E	n
111	1101111	6F	o
112	1110000	70	p
113	1110001	71	q
114	1110010	72	r
115	1110011	73	s
116	1110100	74	t
117	1110101	75	u
118	1110110	76	v
119	1110111	77	w
120	1111000	78	x
121	1111001	79	y
122	1111010	7A	z
123	1111011	7B	{
124	1111100	7C	:
125	1111101	7D	}
126	1111110	7E	~
127	1111111	7F	Δ
128	10000000	80	
129	10000001	81	
130	10000010	82	
131	10000011	83	
132	10000100	84	
133	10000101	85	
134	10000110	86	
135	10000111	87	
136	10001000	88	
137	10001001	89	

138	10001010	8A
139	10001011	8B
140	10001100	8C
141	10001101	8D
142	10001110	8E
143	10001111	8F
144	10010000	90
145	10010001	91
146	10010010	92
147	10010011	93
148	10010100	94
149	10010101	95
150	10010110	96
151	10010111	97
152	10011000	98
153	10011001	99
154	10011010	9A
155	10011011	9B
156	10011100	9C
157	10011101	9D
158	10011110	9E
159	10011111	9F
160	10100000	A0
161	10100001	A1
162	10100010	A2
163	10100011	A3
164	10100100	A4
165	10100101	A5
166	10100110	A6
167	10100111	A7
168	10101000	A8
169	10101001	A9
170	10101010	AA
171	10101011	AB
172	10101100	AC
173	10101101	AD
174	10101110	AE
175	10101111	AF
176	10110000	B0
177	10110001	B1
178	10110010	B2
179	10110011	B3
180	10110100	B4
181	10110101	B5
182	10110110	B6
183	10110111	B7
184	10111000	B8
185	10111001	B9
186	10111010	BA
187	10111011	BB

Decimal	Binary	Hex
188	10111100	BC
189	10111101	BD
190	10111110	BE
191	10111111	BF
192	11000000	C0
193	11000001	C1
194	11000010	C2
195	11000011	C3
196	11000100	C4
197	11000101	C5
198	11000110	C6
199	11000111	C7
200	11001000	C8
201	11001001	C9
202	11001010	CA
203	11001011	CB
204	11001100	CC
205	11001101	CD
206	11001110	CE
207	11001111	CF
208	11010000	D0
209	11010001	D1
210	11010010	D2
211	11010011	D3
212	11010100	D4
213	11010101	D5
214	11010110	D6
215	11010111	D7
216	11011000	D8
217	11011001	D9
218	11011010	DA
219	11011011	DB
220	11011100	DC
221	11011101	DD
222	11011110	DE
223	11011111	DF
224	11100000	E0
225	11100001	E1
226	11100010	E2
227	11100011	E3
228	11100100	E4
229	11100101	E5
230	11100110	E6
231	11100111	E7
232	11101000	E8
233	11101001	E9
234	11101010	EA
235	11101011	EB
236	11101100	EC
237	11101101	ED

238	11101110	EE
239	11101111	EF
240	11110000	F0
241	11110001	F1
242	11110010	F2
243	11110011	F3
244	11110100	F4
245	11110101	F5
246	11110110	F6
247	11110111	F7
248	11111000	F8
249	11111001	F9
250	11111010	FA
251	11111011	FB
252	11111100	FC
253	11111101	FD
254	11111110	FE
255	11111111	FF

Appendix 3
The STE Bus Product Guide

The *STE (IEEE-1000) Bus Product Guide* provides a comprehensive list of STE products. The guide, which is organized into various product categories so that products can be rapidly located within context, includes a wide range of boards (together with brief specifications) and is regularly updated to include new products. Quick reference lists are also included to identify manufacturers and suppliers. The *STE Bus Product Guide* is produced and published by Bob Squirrell Marketing on behalf of the STE Manufacturers and Users Group. The guide currently costs £2.50 plus £1.00 postage and packing (£3.50) from Bob Squirrell Marketing, Unit 6, The Old Mill, Reading Road, Pangbourne, Reading RG8 7HY.

Appendix 4
Commonly used STE digital I/O bus pin connections

```
                    +5 V  | 50 •   • 49 |  +5 V
                   +12 V  | 48 •   • 47 |  -12 V
              ┌─────────────────────────────────────────┐
              │  7     | 46 •   • 45 | 6                │
              │                                         │  Group 4
              │  5     | 44 •   • 43 | 3                │
              │                           ┌─────────────┤
              │  2     | 42 •   • 41 | GND              │
              ├────────────────┬────────────────────────┤
              │  7     | 40 •   • 39 | 6                │
              │                                         │
              │  5     | 38 •   • 37 | 4                │
              │                                         │  Group 3
              │  3     | 36 •   • 35 | 2                │
              │                                         │
              │  1     | 34 •   • 33 | 0                │
              └─────────────────────────────────────────┘
Group 4 Bit  1        | 32 •   • 31 | GND
              ┌─────────────────────────────────────────┐
              │  7     | 30 •   • 29 | 6                │
              │                                         │
              │  5     | 28 •   • 27 | 4                │
              │                                         │  Group 2
              │  3     | 26 •   • 25 | 2                │
              │                                         │
              │  1     | 24 •   • 23 | 0                │
              └─────────────────────────────────────────┘
Group 4 Bit  4        | 22 •   • 21 | GND
              ┌─────────────────────────────────────────┐
              │  7     | 20 •   • 19 | 6                │
              │                                         │
              │  5     | 18 •   • 17 | 4                │
              │                                         │  Group 1
              │  3     | 16 •   • 15 | 2                │
              │                                         │
              │  1     | 14 •   • 13 | 0                │
              └─────────────────────────────────────────┘
Group 4 Bit  0        | 12 •   • 11 | GND
              ┌─────────────────────────────────────────┐
              │  7     | 10 •   •  9 | 6                │
              │                                         │
              │  5     |  8 •   •  7 | 4                │
              │                                         │  Group 0
              │  3     |  6 •   •  5 | 2                │
              │                                         │
              │  1     |  4 •   •  3 | 0                │
              └─────────────────────────────────────────┘
                    GND  |  2 •   •  1 | GND
```

Appendix 5
Commonly used STE analogue I/O bus pin connections

Left	Pin	Pin	Right
+5 V	50	49	+5 V
+12 V	48	47	−12 V
Out chan. 3	46	45	Out chan. 4
Out chan. 1	44	43	Out chan. 0
N/C	42	41	GND
−In chan. 15	40	39	+In chan. 15
−In chan. 14	38	37	+In chan. 14
−In chan. 13	36	35	+In chan. 13
−In chan. 12	34	33	+In chan. 12
GND	32	31	GND
−In chan. 11	30	29	+In chan. 11
−In chan. 10	28	27	+In chan. 10
−In chan. 9	26	25	+In chan. 9
−In chan. 8	24	23	+In chan. 8
GND	22	21	GND
−In chan. 7	20	19	+In chan. 7
−In chan. 6	18	17	+In chan. 6
−In chan. 5	16	15	+In chan. 5
−In chan. 4	14	13	+In chan. 4
GND	12	11	GND
−In chan. 3	10	9	+In chan. 3
−In chan. 2	8	7	+In chan. 2
−In chan. 1	6	5	+In chan. 1
−In chan. 0	4	3	+In chan. 0
GND	2	1	GND

Appendix 6
IEEE-488 pin connections

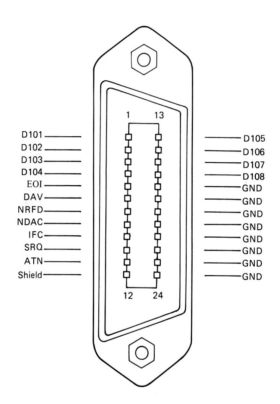

Pin no.	Signal group	Abbreviation	Signal/function
1	Data	D101	Data line 1
2		D102	Data line 2
3		D103	Data line 3
4		D104	Data line 4
5	Management	EOI	End or identify (sent by a talker to indicate that transfer of data is complete)

Pin	Type	Signal	Description
6	Handshake	DAV	Data valid (asserted by a talker to indicate that valid data is present on the bus)
7		NRFD	Not ready for data (asserted by a listener to indicate that it is not ready for data)
8		NDAC	Not data accepted (asserted while data is being accepted by a listener)
9	Management	IFC	Interface clear (asserted by the controller in order to initialize the system in a known state)
10		SRQ	Service request (sent to the controller by a device requiring attention)
11		ATN	Attention (asserted by the controller when placing a command onto the bus)
12		SHIELD	Shield
13	Data	DIO5	Data line 5
14		DIO6	Data line 6
15		DIO7	Data line 7
16		DIO8	Data line 8
17	Management	REN	Remote enable (enables an instrument to be controlled by the bus controller rather than by its own front panel controls)
18		GND	Ground/common
19		GND	Ground/common
20		GND	Ground/common
21		GND	Ground/common
22		GND	Ground/common
23		GND	Ground/common
24		GND	Ground/common

Notes
1 Handshake signals (DAV, NRFD and NDAC) are all active low open-collector and are used in a wired-OR configuration.
2 All other signals are TTL compatible and active low.

Appendix 7
RS-232C pin connections

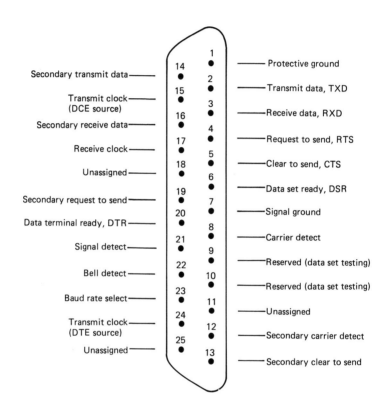

Pin no.	Abbreviation	Direction	Circuit		Function
			CCITT	EIA	
1	FG	–	101	AA	Frame ground
2	TD	To DCE	103	BA	Transmitted data
3	RD	To DTE	104	BB	Received data
4	RTS	To DCE	105	CA	Request to send
5	CTS	To DTE	106	CB	Clear to send
6	DSR	To DTE	107	CC	Data set ready
7	SG	–	102	AB	Signal ground
8	DCD	To DTE	109	CF	Data carrier detect
9		To DTE			Positive DC test voltage

10		To DTE			Negative DC test voltage
11	QM	To DTE	Note 1		Equalizer mode
12	SDCD	To DTE	122	SCF	Secondary data carrier detect
13	SCTS	To DTE	121	SCB	Secondary clear to send
14	STD	To DCE	118	SBA	Secondary transmitted data
	NS	To DCE	Note 1		New synchronization
15	TC	To DTE	114	DB	Transmitter clock
16	SRD	To DTE	119	SBB	Secondary received data
	DCT	To DTE	Note 1		Divided clock transmitter
17	RC	To DTE	115	DD	Receiver clock
18	DCR	To DTE	Note 1		Divided clock receiver
19	SRTS	To DCE	120	DCA	Secondary request to send
20	DTR	To DCE	108.2	CD	Data terminal ready
21	SQ	To DTE	110	CG	Signal quality detect
22	RI	To DTE	125	CE	Ring indicator
23		To DCE	111	CH	Data rate selector
		To DTE	112	CI	Data rate selector
24	TC	To DCE	113	DA	External transmitter clock
25		To DCE	Note 2		Busy

Notes
1 Bell 208A.
2 Bell 113B.

Appendix 8
The STE Manufacturers and Users Group

The STE Manufacturers and Users Group (STEMUG) was formed as an independent non-profit-making forum for the exchange of ideas and information relating to the STE bus (IEEE-1000). The group holds regular meetings at various venues around the country. Membership is currently fixed at £100 per annum and the membership year runs from July to June. Further details of the group can be obtained from the Chairman, STEMUG, PO Box 149, Reading, Berkshire, RG6 3HB.

Appendix 9
Basic commands, compiler directives, functions and statements

The following selection of commands, functions and statements are available in the popular multitasking BASIC compiler known as BASIC-Z80 (B80). This compiler is available in two versions: disk-based and ROM-based. The disk-based version runs under CP/M-Plus. Programs developed on a disk-based system may be programmed into an EPROM to run in a stand alone target system.

Within this appendix, the BASIC language has been divided into the follow subtopics:

Commands	page 271
Compiler directives	page 274
Numeric and special functions	page 275
Statements	page 276
String functions	page 292

Commands

Command	Explanation
CLEAR	Clears all variables, arrays, and procedures.
CLEAR arg1, arg2, arg3	As for CLEAR but arg1 specifies the memory space available for string storage, arg2 specifies the stack space, and arg3 gives the address of the base of usable RAM. *Examples* CLEAR Functionally equivalent to NEW. CLEAR 2000, 512, &8000 Reserves 2000 bytes for string storage, allocates 512 bytes for stack use and sets the bottom of usable RAM to 8000H.
COMP	This command compiles the BASIC program currently resident in memory. The resulting object code may be placed in RAM, saved in EPROM, or saved in HEX format disk file. COMP may also be used with optional arguments.

COMP arg1, arg2, arg3, arg4	If arg1 is specified, the program will be compiled to disk (CP/M-based disk systems) or EPROM (ROM-based systems). The remaining parameters give the start address of the run-time package (arg2), start address of object code (arg3), and RAMTOP (arg4). *Examples* COMP Compiles the resident BASIC program (but does not run it). COMP 'CONTROL', &100, &4000, &C000 Compiles the resident BASIC program to disk in a file called CONTROL.HEX. The run-time package will be located at 100H while the object code will start at 4000H. RAMTOP is set to C000H. COMP '2764', &400, &4000, &D000 Compiles the resident BASIC program to a 2764 EPROM. The run-time package will be located at 400H while the object code will start at 4000H. RAMTOP is set to D000H.
EDIT arg1	EDIT places the first line with a number equal to (or greater than) arg1 into the editing buffer and also lists the line on the console to permit modification. *Example* EDIT 1000 If line 1000 exists, it is listed on the console and placed in the editing buffer. If line 1000 does not exist, the line with the next greater number is listed on the console and placed in the editing buffer.
LIST	Lists the currently resident program at the console. The listing may be paused using CTRL-S (and restarted by any other key), or aborted by CTRL-C. CTRL-P may be used to echo the listing to the printer. Optional parameters may be used in conjunction with the LIST command.
LIST arg1, arg2	arg1 and arg2 specify the first and last line numbers in the listing. *Examples* LIST 100 Only lists line 100. LIST 200 Lists the program from line 200 onwards. LIST 300 Lists from the first line of the program to line 300. LIST 100, 300 Lists the program lines between 100 and 300.
LOAD arg1	Recalls a BASIC program (from disk or EPROM) and loads the program into RAM. An error message is generated if the program file is not found. Arg1 is a file specification.

	Example LOAD 'CONTROL' Searches the directory of the disk placed in the default drive of a CP/M system for a file called CONTROL.BAS. If the file exists, it is loaded into memory. LOAD 'B:MOTOR' Searches the directory of the disk placed in drive B of a CP/M system for a file called MOTOR.BAS. If the file exists, it is loaded into memory. LOAD '2764' Loads a BASIC program (stored as a file on a 2764 EPROM) into RAM. The program is assumed to start at an EPROM address of 00H. LOAD '27128.080' Loads a BASIC program (stored as a file on a 27128 EPROM) into RAM. The program is assumed to start at an EPROM address of 800H. Note that the filename extension is the start address of the program divided by 16.)
MLIST	MLIST can be used to list the addresses of the compiler work areas in order to facilitate development of stand alone programs.
NEW	Clears the BASIC program from memory (including variables). Machine code resident above RAMTOP remains protected.
OLD	This command attempts to recover a program that has been affected by a NEW or CLEAR command. In the case of ROM-based interpreted BASIC, the command attempts to restore a previously resident program that has been deleted by NEW or a system reset. In the case of compiled BASIC running under CP/M, the command attempts to recover the source code by moving the end-of-program marker. In this case, the OLD command is likely to fail and the recovered program will be corrupted if any lines of BASIC have been added in the interim.
RUN	This command compiles the currently resident BASIC source program and, provided the compilation is free from errors, the resulting object code is executed.
SAVE arg1	Stores a BASIC program on disk or EPROM. If an error occurs an error message is generated. Arg1 is a file specification or EPROM type. If an error occurs an error message is generated. Arg1 is a file specification. *Example* SAVE 'CONTROL' Saves the BASIC program as a file on a disk placed in the default drive of a CP/M

	system. The file will be given a default extension of BAS. SAVE 'B:MOTOR' SAVES the BASIC program as a file on a disk placed in drive B of a CP/M system. The file will be given a default extension of BAS. SAVE '2764' SAVEs the BASIC program as a file on a 2764 EPROM starting at an EPROM address of 00H. SAVE '27128.080' SAVEs the BASIC program as a file on a 27128 EPROM starting at an EPROM address of 800H (note that the filename extension is the start address of the program divided by 16).
SYS	In BASIC running under DOS, this command is used to exit to the operating system.
TRACE ON	TRACE ON enables a diagnostic trace facility which prints the line numbers of each line executed as the program runs. TRACE ON is disabled within sections of code that are within scope of an EXCLUDE SCAN statement.
TRACE OFF	TRACE OFF disables the diagnostic trace facility selected by TRACE ON. TRACE OFF is automatically executed if an error occurs or if CTRL-C is entered at the console.
VLIST	VLIST can be used to list the current variables, procedures and functions together with their addresses in memory.

Compiler directives

Directive	Explanation
EXCLUDE SCAN	This compiler directive is used in conjunction with the INCLUDE SCAN directive to identify a block of source code in which no scan calls are required. These sections of the program will run faster since the keyboard will no longer be periodically scanned for control sequences (such as CTRL-C or CTRL-P). Note that, in the absence of EXCLUDE SCAN, the compiler will insert a scan call at the start of each line of source code.
EXT arg1 DEF name	This statement informs the compiler about the type and location of external user-defined functions or procedures (i.e., those occurring in previously compiled sections of code). arg1 is the address of the previously defined procedure while name

enables the procedure to be referenced as if it had been defined in the current source. Name may be followed by one, or more arguments (see DEF name). VLIST may be used to determine the addresses of previously compiled procedures.
Example
EXT &8100 DEF portinit   Informs the compiler of the address (8100H) of a user-defined procedure named portinit. The source code may make subsequent references to portinit without a corresponding DEF portinit statement.

Numeric and special functions

Function	Explanation
ABS (arg1)	Returns the absolute value of the numeric expression given in arg1. *Example* ABS (−1) Returns 1. ABS (2.5) Returns 2.5. ABS (x) Returns 16 when x=16. ABS (x) Returns 16 when x=−16.
DEEK arg1	Returns an integer value corresponding to the two bytes at the address specified in arg1 and arg1+1. The byte at the lowest address is taken as least significant. The address must be in the range 0 to 65535 decimal (0 to FFFFH). *Examples* x%=DEEK (32768) Assigns the numeric value of the 16-bit word stored at a decimal address of 32768 to the integer variable, x%. PRINT DEEK (&01FF) Prints the numeric value of the 16-bit word stored at a hexadecimal address of 1FFH.
INCH	INCH waits until a character is available at the console and then returns its ASCII value. *Example* IF r%=INCH Waits for a character from the console and assigns its ASCII value to the integer variable r%.
INP (arg1)	INP returns the byte at the I/O address specified by arg1 (which must be in the range 0 to 4095). *Example* x%=INP (&80) Reads the byte from an input

	address of 80H and assigns its value to the integer variable x%.
KBD	KBD returns the ASCII code of any key currently depressed. If no key is depressed, KBD returns with zero. Note that, unlike INCH, KBD does not wait for a key depression. *Example* k%=KBD Returns in k% the ASCII value of a currently depressed key. If no key is currently depressed, k% will contain zero.
POS	Returns the current position of the cursor. *Example* z%=POS Returns 0 in z% if the cursor is currently at the extreme left hand side of the console display.
VARPTR (arg1)	Returns the address at which the user variable, procedure or function is stored. When an arg1 specifies an array, the address of the first storage location is returned. If a string variable is specified, VARPTR returns the address of a pointer to the string. *Example* VARPTR (z%) Returns the address at which the integer variable z% is stored. VARPTR (a%) Returns the address of the start of an integer array previously dimensioned using a statement of the form: DIM a%(). VARPTR (a$) Returns the address of the pointer to the string a$.

Statements

Statement	Explanation
BMOVE arg1, arg2, arg3, arg4, arg5	BMOVE can be used to copy a block of memory within the current address space or to/from the STE bus memory. Arguments are used as follows: arg1 Source offset arg2 Source bank number arg3 Destination offset arg4 Destination bank number arg5 Number of bytes to move The offset and bank numbers for the source and destination addresses may be calculated from: Address = (65536 × bank number) + offset Hence a physical address of 21000H is equivalent to bank 2 offset 1000H.

A bank number of −1 indicates that the memory currently in the processor address space is to be used. When both bank numbers are −1 an intelligent block move is performed when source and destination area overlap.

Examples

BMOVE &8000, 0, &D000, 1, &1000   Copies a block of 4096 bytes of memory starting at address 8000H (bank 0) to a destination address of 1D000H (bank 1, offset D000H).

BREAK OFF	Disables monitoring of the console for break (CTRL-C) or pause (CTRL-S) characters.
BREAK ON	Enables monitoring of the console for break (CTRL-C) or pause (CTRL-S) characters.
CALL arg1	Calls the machine code subroutine starting at the address specified by arg1.
CALL arg1, arg2	As for CALL arg1 but arg2 is a pointer to an integer array which contains values to be loaded into the processor's registers. On return, the array is reloaded with the contents of the processor's registers and this mechanism allows passing of parameters to/from machine code modules. The array has four elements as follows: 　0 = accumulator, A 　1 = BC register pair 　2 = DE register pair 　3 = HL register pair *Examples* CALL &C000 Calls the machine code routine at address C000H. The routine should be terminated by a RET instruction. CALL 5, VARPTR (z%) Calls the machine code at address 5 (CP/M BDOS vector) with the register contents supplied by the array z%. The array must have been previously dimensioned by a statement of the form: DIM z%(3). The array is loaded with parameters to be passed using statement of the form: z%(3)=&1000. This example loads 1000H into the HL register pair (H receives 10H whilst L receives 00H) when the machine code routine is entered. After returning to BASIC, the parameter returned in the HL register can be used by statements of the form: If z%(3)=&2000 THEN GOSUB 9000, PRINT 'Result' = ';z%(3) etc.
CLOSE# arg1	Closes a file which has previously been opened for input or output. arg1 is the integer channel specifier previously returned by the operating system as a result of using OPENIN or OPENOUT. *Example* CLOSE# h% Closes the file (and flushes the

	data held in the corresponding buffer) linked to the channel specified by h%. The channel specifier, h%, must have been obtained by previous statements of the form: h% = OPENIN ('filename') or h% = OPENOUT ('filename').
DATA	Specifies items of data to be read by a READ statement. Data items may be numbers or strings separated by commas. Strings may be used without the usual quotation marks PROVIDED they do not contain spaces or commas. *Example* DATA 1.23, 'Bus', 456, 'Control' Specifies four items of data (two numeric and two strings) which will later be READ into BASIC variables.
DEF name	Marks the start of the definition of a named user-defined function or procedure. The function or procedure is subsequently called by name. A user-defined procedure is terminated by ENDPROC or RETURN whilst a user-defined function is terminated by an equality sign followed by the variable (integer, float, or string) to be returned.
DEF name (param1, param2, . . .)	DEF name may also be followed by one or more parameters. When the call is made (by name followed by a list of arguments), the current values of the parameters are saved and the values specified in the argument list are substituted. *Examples* DEF check Marks the start of a user-defined procedure called check. The procedure will be terminated with ENDPROC. DEF ave Marks the start of a user-defined procedure called ave. The function will be terminated by a line of the form: =a% where a% is a variable containing the value to be returned.
DIM name (arg1, arg2, . . .)	The DIM statement is used to reserve memory space for the storage of numeric or string arrays. The name of the array is followed by a list of arguments giving the maximum subscript in each dimension of the array. arg1 thus gives the maximum subscript in dimension 1, arg2 gives the maximum subscript in dimension 2, and so on. Where a maximum subscript is not declared, the default is 10 for each dimension referenced. Note that the maximum number of elements in the array

	will be one greater than the maximum subscript declared and that the first element of the array has a subscript of zero. Hence, an array having a maximum subscript of 10 in a particular dimension can contain a maximum of 11 elements referenced by subscripts of 0, 1, 2, . . . 10. (Most BASICs do not permit the redimensioning of arrays within programs). *Examples* DIM x(20) Dimensions a one-dimensional numeric array, x, with maximum subscript of 20. The array will comprise 21 elements, referenced by x(0), x(1), x(2), . . . x(20). DIM x$(16) Dimensions a one-dimensional string array, x$, with a maximum subscript of 16. The array will comprise 17 elements, referenced by x$(0), x$(1), x$(2), . . . x$(16). DIM y(20, 16) Dimensions a two-dimensional array, y, with maximum subscripts 20 and 16. Elements within the array will be referenced by: y(0, 0) . . . y(0, 16) y(1, 0) . . . y(1, 16) . . . . y(20, 0) . . . y(20, 16) The total number of elements present in the array will thus be 21 × 17 = 357.
DOKE arg1, arg2	This statement sets the 16-bit word stored at address arg1 and arg+1 to the value specified in arg2. arg1 and arg2 are constants or numeric variables in the range 0 to 65535 dec. Note that the low-order byte is stored at arg1 and the high-order byte is stored at arg1+1. *Examples* DOKE &C000, &1000 Stores 1000H at address C000H. The byte at C000H will have the value 10H while the byte at C001H will have the value 00H. DOKE x%, y% Stores FFAAH at address 7FFEH when x%=32766 and y%=65535. The byte values at 7FFEH and 7FFFH will respectively be AAH and FFH.
ELSE	See IF and WHEN.
ENDPROC	ENDPROC is used to mark the end of the definition of a user-defined procedure. ENDPROC has the same effect as RETURN. A user-defined procedure may have multiple ENDPROC or RETURN statements or none at all. Where ENDPROC is omitted, the end of the definition is marked by the end of the program or the start of the next procedure definition.

ENDWHEN	See WHEN.
FOR arg1=arg2 TO arg3	Sets up a loop that will execute one, or more, statements a number of times depending upon the numeric control variable arg1. Note that the loop will be executed faster when arg1 is an integer variable. arg2 and arg3 are numeric constants or numeric variables which respectively specify the initial and final values of the control variable. During execution, the control variable is incremented by 1 on each pass through the loop. The end of the loop must be marked by a corresponding NEXT or NEXT arg1 statement.
FOR arg1=arg2 TO arg3 STEP arg4	As for FOR arg1=arg2 TO arg3 but arg4 specifies the size of increment on each pass through the loop. *Examples* FOR i%=1 TO 10 Constructs a loop that will be executed 10 times with i% as the integer control variable. During execution, i% will take values of 1, 2, 3, . . . 10. Note that the loop will be exited with i% taking a value of 10. The loop must be terminated with NEXT or NEXT i%. FOR j%=0 TO 10 STEP 2 Constructs a loop that will be executed 6 times with j% as the integer control variable. During execution, j% will take values of 0, 2, 4, . . . 10. Note that the loop will be exited with i% taking a value of 10. The loop must be terminated with NEXT or NEXT i%. FOR j%=10 TO 2 STEP −2 Constructs a loop that will be executed 5 times with j% as the integer control variable. During execution, j% will take values of 10, 8, 6, . . . 2. Note that the loop will be exited with j% taking a value of 2. The loop must be terminated with NEXT or NEXT j%. FOR x%=1 TO y% Constructs a loop that will be executed y% times with x% as the integer control variable. During execution, x% will take values of 1, 2, 3, . . . y%. Note that the loop will be exited with x% taking a value of y%.
GET# arg1	Returns the value of a single character (byte) obtained from the input channel specified by the integer, arg1. arg1 can be an integer channel specifier previously returned by the operating system as a result of using an OPENIN statement or can be any currently valid channel recognized by the system. On CP/M systems, for example, channel 0 corresponds to the console while channel 2 corresponds to the CP/M AUX device.

	Examples x%=GET# i% Returns in x% the value of single character (byte) taken from the channel specified by i%. The channel specifier, i%, must have been obtained from a previous statement of the form: i% = OPENIN ('filename') y%=GET# 0 Returns in x% the value of a single character taken from the console (keyboard).
GOTO arg1	Transfers program execution (branches) to the line number specified by arg1. GOTO may be used on its own (unconditional branch) or may follow an IF ... THEN statement (conditional branch). *Examples* GOTO 1000 Transfers program execution to line 1000 (i.e., line 1000 is the next line to be executed). IF x%>1 THEN GOTO 2000 Transfers program execution to line 2000 provided integer x% has the value 2, or more. If this condition is not satisfied, program execution continues with the next higher numbered line.
GOSUB arg1	Transfers program execution (branches) to the subroutine starting at the line number specified by arg1. GOSUB may be used on its own (unconditional subroutine call) or may follow an IF ... THEN statement (conditional subroutine call). In either case, the subroutine must be terminated by a RETURN statement in order that control returns to the next line of the main program. *Examples* GOSUB 1000 Transfers program execution to the subroutine starting at line 1000 (i.e., line 1000 is the next line to be executed). The end of the subroutine is marked by a RETURN statement. IF x%>1 THEN GOSUB 2000 Transfers program execution to the subroutine starting at line 2000 provided integer x% has the value 2, or more. If this condition is not satisfied, program execution continues with the next higher number line. The subroutine must be terminated with a RETURN statement.
If condition THEN statement1	Statement1 is executed if the condition is true (non-zero). If the condition is false (zero) statement1 is ignored and execution continues with the next line of the program.
IF condition THEN statement1 ELSE statement2	Statement1 is executed if the condition is true (non-zero). If the condition is false (zero), statement1 is ignored and statement2 is executed.

	Examples IF x%>10 THEN LET x%=10 Evaluates the expression x%>10 and, if true, x% is made equal to 10 (i.e., if the current value of x% is 11 the expression evaluates to 1 and is therefore true. x% will then be modified so that it is equal to 10). IF t%<65 THEN PRINT 'COOL' ELSE PRINT 'WARM' The value of t% is tested. If t% is currently less than 65, the message 'COOL' is printed on the console otherwise 'WARM' is printed on the console.
INPUT arg1	INPUT causes a question mark to appear at the console and the program then waits for the user to supply an input (terminated by ENTER or RETURN). The input supplied by the user is assigned to arg1 (which may be either a numeric variable or a string variable). Note that the input supplied by the user must be of the correct type.
INPUT prompt; arg1	As for INPUT arg1 but prompt specifies a prompt string to be printed on the console immediately before the question mark.
INPUT arg1, arg2, arg3 . . .	INPUT may also be followed by a list of variables for which values must be supplied by the user. Values contained within a list entered by the user must be separated by commas. Where the ENTER or RETURN key is prematurely pressed, the user will be re-prompted with '?'. *Examples* INPUT v% Generates a '?' on the console and waits for an input supplied by the user (terminated by ENTER or RETURN). The value entered by the user is assigned to the integer variable, v%. Note that an error will occur if the input contains non-numeric characters. INPUT 'Desired value';v% Prints the prompt string 'Desired value' immediately followed by a '?' and waits for a numeric input supplied by the user (terminated by ENTER or RETURN). The value entered by the user is assigned to the integer variable, v%. Note that an error will occur if the input contains non-numeric characters.
INPUT LINE prompt; arg1	As for INPUT prompt; arg1 but the entire input line is placed in the string variable specified by arg1. This statement allows the input string to contain commas which would otherwise be interpreted as separators in a list.
INPUT# arg1	As for INPUT arg1 but input is derived from the

	channel specified by the integer arg1. arg1 can be an integer channel specifier previously returned by the operating system as a result of using an OPENIN statement or can be any currently valid channel recognized by the system. On CP/M systems, for example, channel 0 corresponds to the console while channel 2 corresponds to the CP/M AUX device.
INPUT LINE# arg1	As for INPUT# arg1 but the entire input line is placed in the string variable specified by arg1. This statement allows the input string to contain commas which would otherwise be interpreted as separators in a list. *Examples* x$=INPUT# i% Returns in x$ the string of a single character (byte) taken from the channel specified by i%. The channel specifier, i%, must have been obtained from a previous statement of the form: i% = OPENIN ('filename')
INTERRUPT OFF	This statement disables interrupts until re-enabled by means of an INTERRUPT ON statement. Interrupts are usually disabled during time-critical sections of code. Note that the operating system may use interrupts for console input and in such cases it will be necessary to ensure that an INTERRUPT ON statement appears at the logical end of the program and before making use of statements which involve input from the console.
IS	See WHEN.
LET arg1=arg2	LET is used to assign a value (obtained from arg2) to a variable (arg1). Note that the use of LET is optional in many versions of BASIC. In such cases the following statements are functionally identical: LET arg1=arg2 arg1=arg2 In any event, matching variable types must be used on either side of an assignment statement. *Examples* LET r%=16 Assigns a value of 16 to the integer variable r%. p=3.141593 Assigns the value 3.141593 to the floating point variable p. a=2*s Evaluates the expression 2xs and assigns the result to the floating point variable a. LET e$='Input out of range!' Assigns the string 'Input out of range!' to the string variable e$.

LOCAL arg1, arg2, . . .	LOCAL is used to create variables which are local to the current procedure or function. When a LOCAL statement is encountered within a definition, the current values of the variables specified in the argument list are saved and then restored on return. Note that parameters used in function or procedure declarations are automatically local. *Example* LOCAL i%, v$ Variables i% and v$ will be treated as local within the procedure. The values of i% and v$ in the main body of the program (and in other procedures) will be unaffected by the procedure call.
NEXT	Marks the end of a section of code to be repeated within a FOR . . . NEXT loop.
NEXT arg1	As for NEXT but the loop control variable is stated explicitly. Note that some versions of BASIC require that the loop control variable is always included in a NEXT statement. Others may only require the loop control variable to be stated if it is a floating point type.
ON ERROR name	This statement allows the user to specify a user-defined procedure to be called when an error occurs. The procedure may usefully have a single integer parameter which will be assigned the error number. *Example* ON ERROR alert When an error subsequently occurs, the program calls a user-defined error handling procedure. This procedure definition will start with a statement of the form: DEF alert (errnum%)
ON IN name	The ON IN name statement may be used to redirect subsequent character input via the named user-defined function. Modified versions of the INPUT# and GET# statements are then used for redirected input. *Example* ON IN getchar% Redirects input via the user-defined function, getchar%. Characters to be redirected are subsequently obtained using lines of the form: INPUT##, a$ or GET##, a$ The user-defined function is declared using a statement of the form: DEF getchar%

Statements 285

	and the function definition is terminated by a line of the form: = c% The value in the integer variable, c%, is returned to the INPUT## statement. Note that the function definition must not contain any references to INPUT## or GET## otherwise endless recursion will occur.
ON OUT name	The ON OUT name statement may be used to redirect subsequent character output via the named user-defined procedure. Modified versions of the INPUT# and GET# statements are then used for redirected output. *Example* ON OUT putchar% Redirects output via the user-defined procedure, putchar%. Characters to be redirected are subsequently output using lines of the form: PRINT##, a$ or GET##, a$ The user-defined procedure is declared using a statement of the form: DEF putchar% and the procedure definition is terminated by RETURN. Note that the procedure definition must not contain any references to PRINT## or PUT## otherwise endless recursion will occur.
ON INTERRUPT name	The ON INTERRUPT statement may be used to specify a user-defined procedure that will subsequently be executed whenever an interrupt occurs. The interrupt is vectored via an appropriate jump address in the run-time package (106H and 406H for disk and ROM-based systems respectively). Note that interrupt handling routines must be preceded by EXCLUDE SCAN statements.
ON SCAN name	The ON SCAN statement may be used to specify a user-defined procedure that will subsequently be executed at the beginning of every source line and during keyboard input. The procedure will thus be called repeatedly (though not necessarily at regular intervals) during program execution. ON SCAN may also be used with SWAP TASK to enable multitasking.
OPEN# arg1	Executes the initialization code for the channel number specified by arg1. OPEN# is generally only used in ROM-based systems.

286 Basic commands, compiler directives, functions and statements

	Example OPEN# 2 Initializes channel 2.
OPENIN (arg1)	Opens the file (specified by the string arg1) for input. OPENIN returns an integer channel number which must be used in subsequent INPUT# or GET# statements. *Example* i% = OPENIN ('control.dat') Opens a file for input called 'control.dat' assumed to be present on the default drive in a CP/M system. The channel specifier returned by the system is placed in the integer variable, i%.
OPENOUT (arg1)	Opens the file (specified by the string arg1) for output. OPENOUT returns an integer channel number which must be used in subsequent PRINT# or PUT# statements. *Example* o% = OPENOUT ('control.dat') Opens a file for output called 'control.dat' on the default drive in a CP/M system. The channel specifier returned by the system is placed in the integer variable, o%.
OUT arg1, arg2	The OUT statement sends the integer value arg2 to the port address specified by arg1. arg1 must be an integer in the range 0 to 4096 while arg2 will normally be in the range 0 to 255. Note that if arg2 is greater than 255, its low-order byte will be sent to the port. *Examples* OUT 255, 127 Sends a decimal value of 127 (7FH) to decimal port address 255 (FFH). OUT &FF, &7F Has the same effect as the previous example. OUT p%, x% Sends the low-order byte of the current value of integer variable, x%, to the port specified by the integer variable, p%. If p%=255 and x%=127, this statement will have the same effect as the two previous examples.
POKE arg1, arg2	The POKE statement stores a byte (specified by integer, arg2) at the address specified arg1. arg1 and arg2 should be integers in the ranges 0 to 65535 and 0 to 255 respectively. *Examples* POKE 32768, 127 Stores a decimal value of 127 (7FH) at a decimal memory address of 32768 (8000H). POKE &8000, &7F Has the same effect as the previous example. POKE a%, x% Stores the current value of

integer variable, x%, at the memory address specified by the integer variable, a%. If a% = 32768 and x% = 127, this statement will have the same effect as the two previous examples.

PRINT arg1, arg2, arg3, . . .	The PRINT statement takes a series of arguments (either strings, integers or floating point variables) and outputs their current values to the console display. Items in the argument list must be separated by comma or semi-colon delimiters which are used to determine the position at which printing is resumed. A comma indicates that printing should resume at the next 8-character column position while a semi-colon indicates that printing should continue at the next available print position. The PRINT statement automatically generates a carriage return and line feed (so that the next item printed statement will commence at the extreme left-hand side of the subsequent line) but this provision may be suppressed by terminating the PRINT with a semi-colon. *Examples* PRINT Generates no printed output but generates a carriage return and line feed and thus effectively produces a blank line. PRINT x% Prints the current value of the integer variable x% on the console display and generates a carriage return and line feed so that any subsequent PRINT statement will use the next line of the display. PRINT x%; Prints the current value of the integer variable x% on the console display but does not generate a carriage return and line feed. Any subsequent PRINT statement will use the next available character position. PRINT x%, y% Prints the current values of the integer variables x% and y% on the console display. The two values will be printed on the same line of the display but will be separated. A carriage return and line feed will be generated so that any subsequent PRINT statement will use the next line of the display. PRINT 'Value = '; x Prints the string 'Value = ' immediately followed by the current value of the floating point variable x. A carriage return and line feed is generated so that any subsequent PRINT statement will use the next line of the display. If x has a current value of 123.456 then the statement would produce the following output at the console: Value = 123.456 PRINT 'User name: '; n$ Prints the string 'User

name: ' immediately followed by the current value of the string variable n$. A carriage return and line feed is generated so that any subsequent PRINT statement will use the next line of the display. If n$ has a current value of 'J. Smith' then the statement would produce the following output at the console:
User name: J. Smith
PRINT 'Time used: '; t%; ' sec.'   Prints the string 'Time used: ' immediately followed by the current value of the integer variable t% and the string ' sec.'. A carriage return and line feed is generated so that any subsequent PRINT statement will use the next line of the display. If t% has a current value of 123 then the statement would produce the following output at the console:
Time used: 123 sec.

PRINT# arg1, arg2, arg3, . . .	PRINT# outputs a list of arguments to the channel specified by the integer, arg1. arg1 will have normally been returned by the operating system as a result of a previous OPENOUT statement. PRINT# may also be used to send output to any valid serial channel specified by arg1. *Examples* PRINT# o%, n; Prints the current value of the floating point variable n to the channel specified by the integer, o%. The value of n will thus be saved to a disk file called 'temp.dat' provided the file has been previously opened for output using a statement of the form: o% = OPENOUT ('temp.dat') PRINT#1, n$ Sends the current value of the string n$ to the printer. A carriage return and line feed is generated so that any subsequent PRINT#1 statements will produce output on the next line.

PUT# arg1, arg2	PUT# outputs a character to the channel specified by the integer, arg1. arg1 will normally have been returned by the operating system as a result of a previous OPENOUT statement.
	    PRINT# may also be used to send output to any valid serial channel.
	    Note that PUT# is not restricted to ASCII characters and it may thus be used to generate a binary (non-text) file.
	Example
	PUT# o%, n% Prints the current value of the integer variable n% to the channel specified by the integer, o%. The value of n% will thus be saved to a disk file called 'temp.bin' provided the file has

	been previously opened for output using a statement of the form: o% = OPENOUT ('temp.bin')
READ	Instructs the program to read one or more items of data (contained within a DATA statement) into appropriate BASIC variables. Items in the DATA list may be numbers or strings separated by commas. Strings may be used without the usual quotation marks PROVIDED they do not contain spaces or commas. It is important to ensure that the types of variables used in READ statements match the types of data used in the DATA statements. The position in the data list is held in a data pointer which is automatically incremented as items are read. The data pointer may be altered by means of a RESTORE statement. *Example* READ a, b$, c%, d$ Reads four items of DATA from the data list. Assuming that the data pointer currently points to a line of the form: DATA 1.23, 'Bus', 456, 'Control' then a will receive 1.23, b$ will receive 'Bus', c% will receive 456, and d$ will receive 'Control'.
RESTORE arg1	The RESTORE statement may be used to specify the point within a program from which data is to be read. arg1 specifies the line number at which the DATA list starts. *Example* RESTORE 19000 Sets the data pointer so that the next item of data to be read (using a READ statement) will be taken from line 19000.
REM comment	The REM statements allows the programmer to include comments within a program. Any text on a line following the REM statement will be ignored at run-time. *Example* REM Initializes PORT-B as output The comment following REM generates no executable code – it simply helps the programmer recognize the function of the code module which follows.
REPEAT	REPEAT marks the start of a section of code which will be executed at least once. The end of the section of code must include a matching UNTIL statement. The condition which follows UNTIL is tested and, if found to be true (non-zero) the code will be executed again. If the condition evaluates to false (zero), execution continues with the next higher numbered line.

290 *Basic commands, compiler directives, functions and statements*

	Example
	n% = 1
	REPEAT
	  PRINT n%;
	  n%=n%+1
	UNTIL n%=4
	The above code produces the following output on the console display:
	123
	(Note that the condition is tested at the end of the loop and thus 3 is printed before the loop is exited).

RETURN	RETURN terminates a subroutine (called by a matching GOSUB statement) or may mark the end of the definition of a user-defined procedure. Note that several RETURN statements may be present within a subroutine or procedure definition.
STOP	The STOP statement terminates the current program. In the case of interpreted BASIC, control is returned to the user via the keyboard who may then display and/or modify variables before restarting using CONT. In the case of a stand alone ROM-based system, the program will simply restart.
SWAP TASK	The SWAP TASK statement is used to switch between the current and suspended tasks. The statement is normally used in conjunction with ON SCAN so that tasks are repeatedly switched. *Example* ON SCAN SWAP TASK Switches between the current and suspended tasks at the start of each program line.
TASK name	The TASK statement specifies a user-defined procedure which is to be used as the alternate task in a multitasking application. *Example* TASK clock Specifies the user-defined named clock as the alternate task. The definition of the procedure will be declared with a line of the form: DEF clock and it will be terminated with a RETURN statement.
TRACE ON	TRACE ON enables a diagnostic trace facility which prints the line numbers of each line executed as the program runs. TRACE ON is disabled within sections of code that are within scope of an EXCLUDE SCAN statement.
TRACE OFF	TRACE ON disables the diagnostic trace facility

	selected by TRACE ON. TRACE OFF is automatically executed if an error occurs or if CTRL-C is entered at the console.
WEND	WEND marks the end of a section of code following a WHILE statement.

WHEN arg1	The WHEN statement provides a means of selecting one of several options depending upon the outcome of the expression in arg1. WHEN is used in conjunction with IS and ENDWHEN. arg1 is evaluated and truncated to an integer. The result is then used in a series of comparisons. The statement following the first comparison found to be true is then executed. Control then passes to the next higher number line following the ENDWHEN statement.
	
	If no condition is found to be true, the optional ELSE statement is executed. Note that the WHEN statement may be effectively used as a multi-line IF statement, avoiding the use of GOTO.
	Example
	<pre>WHEN n%-68
	    IS < -2 THEN PRINT "Too cold"
	    IS = 0 THEN PRINT "Exact temperature"
	    IS > 2 THEN PRINT "Too hot"
	    ELSE PRINT "Temperature in range"</pre>If n% is less than 66, 'Too cold' is printed. If n% is greater than 70, 'Too hot' is printed. If n% is 68, 'Exact temperature' is printed. If none of the foregoing is true n% has a value of 66, 67, 69, or 70 in which case 'Temperature in range' is printed.

WHILE arg1	WHILE marks the start of a section of code which will be executed depending upon the condition in arg1. The end of the section of code is marked by a matching WEND statement. The condition which follows WHEN is tested on each pass through the loop. The code within the loop will continue to be executed as long as the condition evaluates to true (non-zero). If the condition evaluates to false (zero), program execution continues with the next higher numbered line following the matching WEND statement.
	Example
	<pre>n% = 1
	WHILE n% < 4
	    PRINT n%;
	    n%=n%+1
	    WEND</pre>

The above code produces the following output on the console display:
123
Note that, since the condition is tested at the beginning of the loop, the loop is no longer executed when x%=4.

String functions

Function	Explanation
ASC (arg1)	Returns the ASCII value of the first character of the string given in arg1. *Examples* ASC ('A') Returns 65. ASC ('a') Returns 97. ASC (X$) Returns 65 when X$='A' ASC (X$) Returns 42 when X$='*' ASC (X$) Returns 65 when X$='ASCII'
CHR$ (arg1)	Returns a single character string corresponding to the ASCII value given in arg1. *Examples* CHR$ (65) Returns 'A'. CHR$ (&41) Returns 'A'. CHR$ (x%) Returns '*' when x%=42.
HEX$ (arg1)	Returns a string containing the hexadecimal representation of arg1. *Example* HEX$ (255) Returns 'FF'.
INSTR (arg1, arg2)	Searches for the first occurrence of the string specified by arg2 within the string specified by arg1. If a match is found, the position (in terms of characters from the the start of arg1) is returned. If no match is found, INSTR returns with zero. *Example* n%=INSTR (r$,' ') Searches the string variable r$ for the first occurrence of a space (ASCII 32). The position of the first space detected is returned in the integer n%.
LEFT$ (arg1, arg2)	Returns a substring of arg2 characters taken from the string specified by arg1 starting at the left. *Example* LEFT$ ('CONTROL',3) Returns 'CON'. LEFT$ (X$, Y%) Returns 'MOTO' when X$='MOTOR' and Y%=4.

LEN (arg1)	Returns the length of the string specified in arg1. *Example* LEN ('CONTROL') Returns 7. LEN (x$) Returns 3 when x$='bus'.
MID$ (arg1, arg2, arg3)	Returns a substring of arg3 characters taken from the string specified by arg1 starting with the character position specified by arg2. *Example* MID$ ('Control', 3, 2) Returns 'tr'. MID$ (x$, y%, z%) Return 's con' when x$='Bus control', y%=2, and z%=5.
RIGHT$ (arg1, arg2)	Returns a substring of arg2 characters taken from the string specified by arg1 starting at the right. *Example* RIGHT$ ('Control',3) Returns 'rot'. RIGHT$ (x$, y%) Returns 'otor' when x$='Motor' and y%=4.
STR$ (arg1)	Returns a string equivalent to the numeric value of arg1. *Examples* STR$ (123) Returns '123'. STR$ (x%) Returns '123' when x%=123.
VAL (arg1)	Returns the numeric value of the string specified by arg1. Trailing non-numeric characters are ignored. *Examples* VAL ('123') Returns 123. VAL ('6800 CPU') Returns 6800. VAL (x$) Returns 1 when x$='1.23'.

Appendix 10
Basic Z80 procedure library

```
9999  REM ----------------------------------------
10000 REM    *** General Purpose Procedures ***
10099 REM ----------------------------------------
10100 REM Delay routine
10110 DEF delay(lim%)
10120    LOCAL c%
10130    FOR c%=0 TO lim%: NEXT c%
10140 ENDPROC
10199 REM ----------------------------------------
10200 REM Clear screen
10210 DEF clear_screen
10220    LOCAL n%
10230    FOR n%=1 TO 25
10240       PRINT
10250    NEXT n%
10260 ENDPROC
10299 REM ----------------------------------------
10300 REM Scroll a given number of lines
10310 DEF scroll(lin%)
10320    LOCAL n%
10330    FOR n%=1 TO lin%
10340       PRINT
10350    NEXT n%
10360    ENDPROC
10399 REM ----------------------------------------
10400 REM Console bleep
10410 DEF bleep
10420    PRINT CHR$(7);
10430 ENDPROC
10499 REM ----------------------------------------
10500 REM Alarm signal
10510 DEF alarm
10520    LOCAL n%
10530    FOR n%=1 TO 5
10540       bleep
10545       FOR s%=1 TO 2000: NEXT
10550    NEXT n%
10560 ENDPROC
10599 REM ----------------------------------------
10600 REM Wait for space key
10610 DEF wait_space
10620    PRINT "Press [SPACE] to continue ..."
10630    IF KBD<>32 THEN GOTO 10630
10640 ENDPROC
10699 REM ----------------------------------------
10700 REM Shorten string to given length
10710 DEF shorten(long$,length%)
10720    string$=LEFT$(long$,length%)
10730 ENDPROC
10799 REM ----------------------------------------
10800 REM Print lines of given character
10810 DEF pr_line(lim%,max%,char%)
10820    FOR lin%=1 TO lim%
10830       FOR col%=1 TO max%
10840          PRINT CHR$(char%);
10850       NEXT col%
10855       PRINT
10860    NEXT lin%
10870 ENDPROC
10899 REM ----------------------------------------
```

```
10900 REM Get confirmation flag from user
10910 DEF confirm
10920   LOCAL k%
10930   f%=-1
10940   PRINT "Are you sure ? (Y/N)"
10945   scroll(4)
10950   REPEAT
10960   k%=KBD
10970   IF k%=&59 OR k%=&79 THEN f%=1
10980   IF k%=&4E OR k%=&6E THEN f%=0
10990   UNTIL f%+1
10995 ENDPROC
10999 REM ----------------------------------------
11000 REM General purpose numerical input
11010 DEF number_in(min%,max%)
11020   bleep
11030   PRINT prompt$;" ";
11035   INPUT n$
11040   n%=VAL(n$)
11050   IF n%<=max% AND n%>=min% THEN ENDPROC
11055   PRINT
11060   PRINT "Value outside permissible range (";
11061   PRINT min%;" to ";max%;") !"
11065   PRINT
11070   alarm
11080   GOTO 11030
11090 ENDPROC
11099 REM ----------------------------------------
11100 REM General purpose string input
11110 DEF string_in(length%,default$)
11120    bleep
11130    PRINT prompt$;" ";
11140    INPUT LINE r$
11150    IF r$="" THEN r$=default$
11160    shorten(r$,length%)
11170 ENDPROC
11199 REM ----------------------------------------
11200 REM Upper case converter
11210 DEF upper(str$)
11220   LOCAL u%,v%
11225   string$=""
11230   FOR u%=0 TO LEN(str$)
11240      v%=ASC(MID$(str$,u%,1))
11250       IF v%>96 THEN v%=v%-32
11260      string$=string$+CHR$(v%)
11270   NEXT u%
11280 ENDPROC
11299 REM ----------------------------------------
11300 REM Lower case converter
11310   DEF lower(str$)
11320   LOCAL u%,v%
11325   string$=""
11330   FOR u%=0 TO LEN(str$)
11340      v%=ASC(MID$(str$,u%,1))
11350       IF v%>64 AND v%<91 THEN v%=v%+32
11360      string$=string$+CHR$(v%)
11370   NEXT u%
11380 ENDPROC
11399 REM ----------------------------------------
```

Appendix 11

An overview of STEbus from the STE Manufacturers and Users Group

The Origins of STEbus

STEbus was created in 1982 when the Institute of Electrical and Electronic Engineers in the US, the international standard makers for bus systems, set up a working group to develop it. The term STEbus was coined from visions of a Eurocard version of the STD bus, the then dominant 8-bit standard.

The end result was, however, very different. Taking full account of the latest trends in processor and system design, the working group created a totally new low-cost 8-bit bus to meet the needs of today's designers – and those of tomorrow.

The major buses of the early eighties were closely associated with particular manufacturers and had become, through their popularity, essentially de facto standards. STEbus broke new ground as the first bus system to be created independently of commercial interests. It was designed by engineers for engineers.

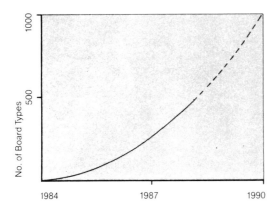

Figure 1 *In less than four years the STEbus has become a major international bus standard. There are now over 700 board types available to system integrators and over 50 suppliers of STEbus products worldwide.*

Since the specification was agreed in 1984, the number of STEbus boards and suppliers has mushroomed (Figure 1). The STEbus now has a major following and is easily the fastest growing 8-bit bus. Support for

STEbus is highest in the UK where most of the pioneering work was done, but the bus is also rapidly gaining popularity in both Europe and the US.

Why another eight-bit bus?

By the early eighties, backplane buses such as STD which were developed in the mid-seventies were already showing their age. Such trends as faster processors, wider addressing ranges, CMOS and multiprocessing were slowly making them obsolete. Only limited upgrades were possible while maintaining compatibility with boards produced to the original standard. STEbus's creators saw the need for a next generation bus.

Though processors were gradually moving up in complexity from 8- to 16- to 32-bits, a conscious decision was made during STEbus's development to restrict data transfers to 8 bits, and 8 bits only. There was one simple reason for this – cost. The added complexity of allowing wider data paths would have significantly increased the cost of each STEbus board and reduced the board area available for useful functions.

Within its class, STEbus nonetheless offers high performance, allowing data to be transferred at over 5 Mbytes/sec. Since 8-bit versions of

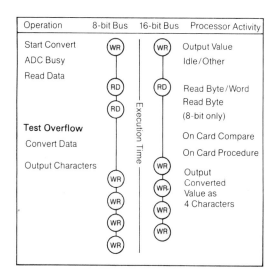

Figure 2 *There can be little advantage in using wide buses for I/O in control and instrumentation systems. In the data path for an average speed 12-bit ADC to a logging device, the only gain apparent from using the 16-bit bus is a reduction in time of one Read cycle.*

popular 16- and even 32-bit processors are widely available, the STEbus user is not restricted to older processor technology.

In such fields as industrial control, 8-bit computing is, and will remain, perfectly suitable for most systems (Figure 2). Though engineers who want powerful number-crunching systems would be better advised to look to a 32-bit bus standards such as VME or Futurebus, they could also benefit from STEbus as an I/O bus (Figure 3).

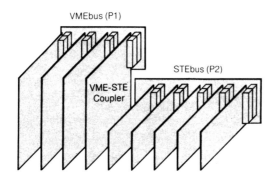

Figure 3 Since VME and STEbus are both Eurocard buses they can be readily combined in the same rack. A VME/STEbus coupler board links the two buses together.

It is easy for a 16-bit or 32-bit bus to become I/O bound since most interface standards to industrial computers are byte organized – SASI, SCSI, IEEE-488 and RS-232 for example. Data must often be transferred between a peripheral device and the CPU a byte at a time, wasting ¾ of the bus bandwidth in a 32-bit system. In contrast, STEbus can make full use of its bus.

STEbus's 8-bit bus also avoids the problems faced by 32-bit systems in coping with the different ways various microprocessors justify data bytes. Byte swapping logic is unnecessary and this, coupled with far fewer transceiver and control components, leads to an interface that is easy to implement and inexpensive.

The simplicity of the interface means significant savings over 16-bit and 32-bit systems in both the design time and production cost of any custom boards that are made. It also allows a relatively small board size. STEbus is defined as single or double Eurocard, though single height boards are more common.

A small board size also helps keep costs low through less expensive packaging and reduced granularity. Designers are more likely to be able to configure precisely the functions they require if there are fewer per board.

Why Eurocard?

Almost all the more recent buses have adopted the Eurocard form factor, and this standardization on board sizes is a major boon to the system designer. Eurocard mounting hardware is readily and cheaply available from many sources, as are such items as prototyping boards, power supplies, modern cards and networking hardware.

Apart from offering the widest choice, Eurocard allows engineers to maximize the use of enclosure space and ancillaries such as power supplies while reducing costs by including a variety of functions, whether related or not, within the same box. The unique connectors and board sizes used by the older buses have become their Achilles heel as more engineers insist on using the Eurocard format.

How long will STEbus last?

Recognizing that a backplane bus represents a major investment, great care was taken in developing the specification to give STEbus as long as life as possible. Building-in such advanced features as processor independence and multiprocessor capability were key to this philosophy, as was the generous allocation of address and I/O space.

Just as fundamental, however, was the choice of data transfer protocol. Older buses were commonly synchronous, so their maximum performance was limited at their inception, allowing little room for speed improvements as device technology improved. Like other modern buses such as VMEbus and Futurebus, STEbus overcomes this problem by employing an asynchronous handshake protocol.

With such a system, the speed of data transfer is governed by the slowest participating board, and not by timing figures in the specification. This allows the possibility of performance improvements as and when faster devices can be used. Also, STEbus's protocol does not unduly favour any particular processor family, thereby giving the designer a wide choice of CPU and facilitating the mixing of processor types in a multiprocessor system.

How does STEbus handle multiple masters?

Multiprocessing can allow designers to boost system performance without recourse to costly 32-bit microprocessors and can also simplify the development effort. If, for example, one processor in a system is made to handle communications, another the real-world interfacing and a third overall system control, the system can readily be partitioned into functional modules, facilitating a team effort on both hardware and software development.

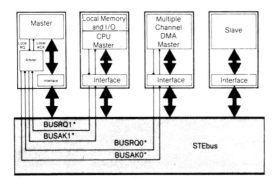

Figure 4 *An example of a multiple master STEbus implementation showing the use of the bus request and request grant lines. An asterisk after the signal name indicates that the signal is active-low.*

Providing multi-master facilities in a bus can incur significant penalties in the complexity and cost of the bus interface. Hence STEbus caters for only three bus masters, an acceptable compromise that keeps bus arbitration simple and inexpensive – it can typically be implemented in a single programmable logic array (PAL).

STEbus defines two types of bus master. A 'Default Master' is allocated the bus by the Arbiter whenever the bus is not in use by another master. Typically the Default Master would be a board that spends most of its time communicating with other parts of the system and hence for which short bus access times are important.

A 'Potential Master' must request use of the bus from the Arbiter by asserting one of the two Bus Request (BUSRQn*) lines. The Arbiter can allocate control according to a variety of schemes, round robin or priority for example. It hands the bus over to a master by asserting one of the two Request Grant (BUSAKn*) lines.

Two modes of operation are possible. In Release-When-Done mode, the current master retains control of the bus until it completes all its transfers. In the other option, Release-On-Request, the master only relinquishes the bus when another master requests it.

To ensure that no spurious bus requests are generated in hostile environments, but requests on the BUSRQn* lines are level rather than edge triggered. Also, the current master asserts its BUSRQn* line continuously while it is using the bus. This is a useful feature in cases where data must be transferred in a continuous stream and could be lost if another master were allocated the bus during the transfer.

How does STEbus handle Interrupts?

The STEbus allows interrupts to be processed in several ways. At the simplest level, for such interrupts as Power Fail, no acknowledge is necessary and an ATNRQn*line is asserted by the interrupting module. For 'common' interrupts, the interrupting module is acknowledged by a read or write operation on one of its registers.

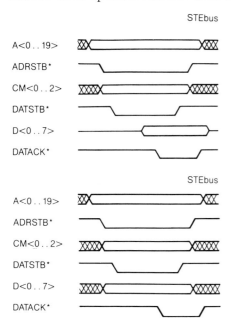

Figure 5 *Read and Write cycles for the STEbus. In the Read cycle the address bus is driven by the master and qualified by ADRSTB*. The command bus is then driven by the master and qualified by DATSTB*. Finally, the slave drives the data bus and the DATACK* handshake line when the data is valid. The Write cycle proceeds in similar fashion.*

Bus-vectored interrupts are the most powerful. In these, an interrupt handler uses the bus's command modifier lines to indicate an acknowledge cycle and puts the encoded ATNRQn* line number onto the address bus as a 3-bit address. The interrupting module can then put an interrupt vector onto the bus which the handler reads in the acknowledge cycle.

How is priority determined?

Unlike a number of other buses that use daisychaining and hence impose device dependent timing constraints, STEbus allows boards to be placed anywhere along the backplane. Priority is determined by which of the eight Attention Request (ATNRQn*) lines a board asserts. ATNRQ0* holds highest priority, ATNRQ7* the least. The position independence of STEbus boards facilitates system building and servicing and is a far more flexible solution for generating interrupts.

STEbus as a secondary bus

The STEbus is widely used to free high performance bus systems from I/O intensive tasks. Considerable enhancements to 32-bit system performance can be achieved by offloading the burden of I/O processing from the main processor and bus. Data flow between a system and external device is often slow and incurs a significant interrupt servicing overhead.

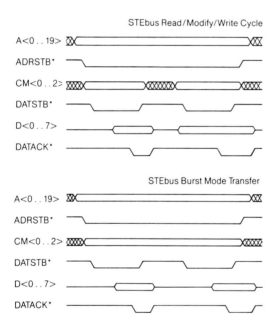

Figure 6 *The STEbus Read/Modify/Write cycle and Burst Mode Transfer. In the Read/Modify/Write cycle, the master holds the address lines stable but changes the command bus. Two handshake sequences – first the read cycle then the write cycle – are performed. In the Burst Mode Transfer, the slave performs the first access at the address supplied by the master. The master holds the address constant for subsequent cycles while the slave increments the address.*

An intelligent subsystem running a real-time executive separate from the rest of the system can organize data into pages or packets before interrupting the main processor's flow. Similarly, large packets of data can be handed to the I/O subsystem from the main processor for subsequent output.

A further advantage of using STEbus as an I/O bus is that many functions on STEbus boards are not available on the larger boards used by 32-bit buses – either because they are only provided as one of many functions on multifunction boards or because the market does not warrant the considerably higher development and production costs.

While STEbus can be used as an I/O subsystem bus with a variety of high performance buses, the combination of STEbus and VME is a natural choice for many engineers. The VMEbus is the most popular of the 32-bit buses currently available and offers the widest range of board types and manufacturers. It is also highly compatible with STEbus (Figure 3).

This is no coincidence. Both buses were developed at about the same time and the two IEEE working groups collaborated to allow the buses to work readily together. VMEbus uses the same power supply voltages, power-fall signalling and system clock and reset as STEbus. Also, STEbus was defined to use only the outer two rows of the two part DIN 41612 connector specified by Eurocard, leaving the centre now free for VMEbus 32-bit extension signals.

What about future upgrades?

The in-built compatibilities between STEbus and VME reflect an unstated objective of the STEbus working group – that both physical and conceptual upwards migration paths from STEbus to VME should be provided to allow systems engineers to maintain their investment in STEbus when developing more powerful next generation systems.

Coupler boards which link STEbus bus subsystems to VME are available. Using these, a large portion of an existing STEbus system can literally be grafted onto the new VMEbus based system. With an intelligent interface, the existing software can also be used.

Can STEbus operate in harsh environments?

The STEbus is designed to be as tolerant as possible in harsh environments. By adopting the Eurocard format, STEbus uses the highly reliable two-part DIN 41612 connector, a great improvement on the direct edge connectors used by older bus standards. All signal lines are buffered and the backplane properly terminated.

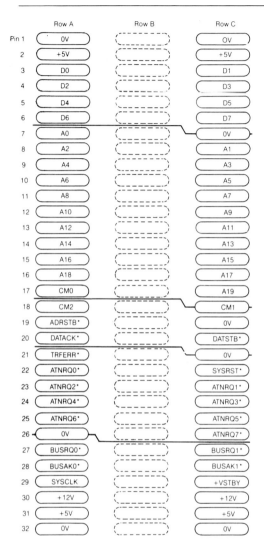

Figure 7 The 64-signal STEbus pin-out defined on rows A and C of a DIN 41612 connector. Address lines A0–19 provide 1 Mbyte of main memory addressing. Depending on the cycle, A0–11 are used to address the 4 kbytes of I/O space and A0–2 provide a 3/bit acknowledge address. Lines D0–7 are the 8-bit data bus. ADRSTB* and DATSTB* are address and data strobes. Lines CM0–2 define the type of bus cycle in progress (whether memory or I/O read or write or an acknowledge). Request lines BUSRQ0,1* are used by temporary masters. DATACK* is asserted when a master accepts data (on a read cycle) or when data is valid (on a write-cycle). TRFERR* is used if data from a slave is incorrect. Signals ATNRQ0–7* are attention request lines and SYSCLK* and SYSRST* are for a 16 MHz clock and reset functions. The remaining lines are for power and fully distributed grounds.

A system error signal is also provided to further ensure the integrity of data transfers. This is asserted by the system controller if an acknowledge is not returned within a given time or by a slave if a local error occurs during a transfer. Crosstalk is minimized by the careful layout of signal lines on the DIN 41612 connector.

Who is using STEbus?

The STEbus has become widely used across industry – from communications and laboratory measurements through to machine and process control. In many cases, it is taking over the role of STD as a low-cost industrial bus. STEbus is also creating new markets. Companies are, for example, finding it better to build their own programmable logic controllers from STEbus boards rather than use inflexible ready-made PLC's.

With its advanced features and ability to use the latest processors, the STEbus is also proving popular in less demanding information processing applications in preference to a higher performance 32-bit bus such as VME. Generally, if STEbus can do the job, and system engineers resist the temptation to overspecify, they can build their systems at a fraction of the cost.

Is the STEbus specification fully approved?

Yes. In December 1987, STEbus finally gained full IEEE approval. It is now known as IEEE 1000–1987 standard.

The advantages of STEbus

- Designed by engineers for engineers
- Manufacturer independent – widest processor range of any 8-bit bus
- Performance into the future – asynchronous, non-multiplexed data transfers at over 5 Mbytes/sec
- Full 1 Mbyte addressing range
- Extensive I/O capability – up to 4 kbytes of I/O space
- Position independent, non-daisychained bus
- Designed as a low-cost – 8-bit bus embracing Eurocard standard
- Multiprocessing capability
- High speed burst transfer mode
- Eight attention request lines
- Vectored or non-vectored interrupts

- Interrupt acknowledge cycle
- Read modify-write cycle
- Designed for high data integrity. Fully buffered signals, terminated backplane
- Now approved as a full standard by the Institute of Electrical and Electronic Engineers

For information on the supply of literature, assistance and specifications, contact:

Bob Squirrell
Chairman
STE Manufacturers and Users Group
PO Box 149
Reading
England
RG6 3HB

Manufacturers and
Users Group Hotline
Tel: (07357) 4976

Tel: (07357) 5445
Fax: (07357) 5185

Copies of the IEEE 1000 (STEbus) specification can also be obtained in the USA from:

The IEEE Computer Society
PO Box 80452
Worldway Postal Centre
Los Angeles
CA 90080
USA

The IEEE will levy a charge to cover the cost of the document and despatch costs.

Glossary of bus terms

Active high Property of a signal which is asserted in the high (logic 1) state.
Active low Property of a signal which is asserted in the low (logic 0) state.
Address A reference to the location of data in memory or within I/O space. Also, the value represented by the state of the address lines while ADRSTB* is asserted.
Address bus The set of lines used to transmit address information. The STE bus has twenty address lines (A0 to A19) and these are capable of addressing a total address range of 1 megabyte. STE I/O operations, on the other hand, make use of address lines A0 to A11 thus the STE bus has a 4 kilobyte I/O range.
Address decoder A hardware device (often a single integrated circuit) which provides chip select or chip enable signals from address patterns which appear on an address bus.
Address selection The process of selecting a specific address or range of addresses on a slave I/O or memory board. Links are usually provided so that such cards may be configured to appropriate I/O or memory addresses.
Addressed board A board that recognizes its address while ADRSTB* is active.
Arbiter A functional module that accepts bus requests and grants control of the data transfer bus to one master at a time.
Arbitration The process of assigning control of the data transfer bus between competing masters.
Asserted The logical true state of a particular signal.
Backplane A printed circuit board on which connectors are mounted at regular intervals. Boards are inserted into the connectors at right angles to the backplane.
Bit A contraction of 'binary digit'; a single digit in a binary number.

Block transfer A sequence of data transfers, in the same direction, which occur during a single bus transaction.

Board A printed circuit board which complies with the physical and electrical specification for a particular bus system and which contains a functional module such as a processor, memory, or I/O device.

Bus A path for signals having some common function. Most microprocessor systems have three distinct buses: an address bus, data bus and control bus. The STE bus contains the full range of signals to facilitate asynchronous data transfer between functional modules.

Bus cycle A sequence of events in a bus system which constitute a complete operation (such as transferring data from a bus master to a bus slave).

Byte A group of eight bits which is operated on as a unit.

Card A printed circuit board which complies with the physical and electrical specification for a particular bus system and which contains a functional module such as a processor or I/O subsystem.

Central processing unit The part of a computer that decodes instructions and controls the other hardware elements of the system. The CPU comprises a control unit, arithmetic/logic unit and internal storage. A microprocessor fulfills the functions of a CPU.

Chip The term commonly used to describe an integrated circuit.

Compatibility The degree to which the boards in a bus system may be interconnected and used without further modification.

Daisy-chain A method of propagating a signal from board to board within a bus system. The signal emanates from a board placed in one slot and propagates through each of the other boards connected to slots accorded a higher priority by virtue of their position.

Data General term used to describe numbers, letters and symbols present with a computer system. Also, the value represented by the state of the data lines when DATSTB* is asserted.

Data bus The set of electrical conductors which carries data between the different functional modules (for example bus masters and bus slaves) of bus system. Also known as the *data transfer bus*.

Data transfer bus cycle A sequence of operations that result in the transfer of data between a master and slave.

Firmware A program (software) stored in read-only memory (ROM).

Functional module A collection of electronic circuitry which resides on a board or card in a bus system and which works together to accomplish a task.

Handshake An interlocked sequence of signals between interconnected boards in which each board waits for an acknowledgement of its previous signal before proceeding.

Hardware The physical components of a bus system.

High state The more positive of the two voltage levels used to represent binary logic states. A high state (logic 1) is generally represented by a voltage in the range 2 V to 5 V.

Input/output (I/O) Devices and lines used to transfer information to and from external (peripheral) devices.

Input port A circuit that allows a functional module to accept signals from external devices.

Integrated circuit An electronic circuit fabricated on a single wafer (chip) and packaged as a single component.

Intelligent slave A slave which incorporates a processor (or equivalent logic) but which is incapable of controlling the bus (i.e., cannot act as a bus master).

Interface A shared boundary between two or more systems, or between two or more elements within a system.

Interface system The functional elements required for unambiguous communication between two or more devices. Typical elements include: driver and receiver circuitry, signal line descriptions, timing and control conventions, communication protocols, and functional logic circuits.

Low state The more negative of the two voltage levels used to represent the binary logic states. A low state (logic 0) is generally represented by a voltage in the range 0 V to 0.8 V.

Master A board fitted with a processor (or equivalent logic) which is capable of controlling the bus. Up to three masters may be present in an STE bus system.

Memory That part of a bus system into which information can be placed and later retrieved. Storage and memory are interchangeable terms.

Microprocessor A central processing unit fabricated on a single chip.

Noise Any unwanted signal component which may appear superimposed on a wanted signal.

Operating system A control program which provides a low-level interface with the system hardware. The operating system thus frees the programmer from the need to produce hardware specific I/O routines (for example those associated with disk filing).

Output port A circuit that allows a functional module to output signals to external devices (peripherals).

Peripheral An external hardware device whose activity is under the control of the bus system.

Processor The name given to a functional module which is capable of controlling the other functional elements of a bus system. A bus master contains a processor (or equivalent logic).

Protocol The signalling rules used to convey information or commands between boards connected to a bus.

RAM (random access memory) Usually used to mean semiconductor read/write memory. Strictly speaking, ROM devices are also random access.

Random access An access method in which each word can be retrieved in the same amount of time (i.e., the storage locations can be accessed in any desired order).

Read The process of transferring information to a bus master from memory or I/O.

Release The action of a controlling device in ceasing to hold a signal line in an asserted state.

ROM (read-only memory) A memory device which is permanently programmed. Erasable-programmable read only memory (EPROM) devices are popular for storage of programs and data in stand alone bus applications and can be erased under ultra-violet light to permit reprogramming.

Sequence An individual bus transaction comprising one or more transfers.

Settling time The time taken for a signal line to settle to a defined logical state when making a transition from one state to another.

Signal The information conveyed by an electrical quantity present in one of the conductors of a bus system.

Signal level The relative magnitude of a signal when considered in relation to an arbitrary reference (usually expressed in volts, V).

Signal line One of a set of signal conductors in a bus system used to transfer data or control information.

Signal parameter That element of an electrical quantity whose values or sequence of values is used to convey information.

Slave A board which is capable of accepting and executing commands issued by a master.

System A set of interconnected boards which achieves a specific objective by the performance of designated functions.

Tenure The time during which a master has control of the bus.

Transaction The combination of data transfer sequences controlled by a master during a single bus tenure.

Transfer The movement of a single byte of data from the current master to an addressed slave(s) or from the addressed slave to the master.

Visual display unit (VDU) An output device (usually based on a cathode ray tube) on which text and/or graphics can be displayed. A VDU is normally fitted with an integral keyboard in which case it is sometimes referred to as a console.

Write The process of transferring information from a master to memory or I/O.

Bibliography

Control systems, instrumentation and measurement

Berk, A. A., *Micros in Process and Product Control*, Collins. ISBN 0 00 383296 1.
  A useful introduction to microcomputer-based control systems and eminently suitable for those not possessing a formal microelectronics background.

Money, S. A., *Microprocessors in Instrumentation and Control*, Collins. ISBN 0 00 383041 1.
  Deals in a practical manner with the application of microprocessors in the field of control and instrumentation. Assumes that the reader has a basic understanding of the principles of digital electronics.

Cassell, Douglas, *Microcomputers and Modern Control Engineering*, Cassell 1983. ISBN 0 8359 4365 8.
  A comprehensive guide to the principles and practice of modern control engineering.

Electronic circuits

Horowitz, Paul and Hill, Winfield, *The Art of Electronics*, Cambridge University Press 1980. ISBNs 0 521 23151 (hardback); 0521 29837 (paperback).
  A comprehensive guide to the design of electronic circuits. The book is eminently readable and adopts a practical approach. Recommended reading for those having a limited electronic background.

Tooley, Michael, *Electronic Circuits Handbook – Design, testing and construction*, Heinemann Professional Publishing 1987. ISBN 0 434 91968 3.
  Provides readers with a unique collection of practical circuits together

with supporting information so that working circuits can be produced in the shortest possible time and without recourse to theoretical texts.

Digital techniques

Lancaster, Don, *The TTL Cookbook*, Howard Sams. ISBN 0 672 21035 5.

An invaluable collection of hints, tips, facts and figures covering all facets of TTL. A selection of the most popular TTL devices is discussed in some detail (together with pinouts for each device). The book also has a useful section on timers.

Lancaster, Don, *The CMOS Cookbook*, Howard Sams. ISBN 0 672 21398 2.

This book is similar to its TTL counterpart and makes equally good reading.

Texas Instruments Staff, *The TTL Data Book for Design Engineers*, Texas Instruments (Europe). ISBN 3 88078 034 X.

The definitive text covering all types of TTL device. Includes electrical characteristics and pin connecting data.

Tooley, Michael, *Practical Digital Electronics Handbook*, PC Publishing. ISBN 1 870775 00 7.

A practical guide to digital circuits and microprocessor-based systems.

Towers, T. D., *The Towers' International Digital IC Selector*, Foulsham. ISBN 0 572 01179.

Provides abridged data and pin connecting information for over 13,000 digital integrated circuits. Appendices provide useful reference information on IC logic types and codings, package outlines, pinouts, manufacturers' codings, manufacturers' proprietary 'house' codings, abbreviations and a glossary.

Tooley, Michael, *The Computer Engineer's Pocket Book*, Heinemann Professional Publishing 1986. ISBN 0 434 91967.

A compendium of facts, figures, circuits and data and includes TTL and CMOS pinouts, logic gate characteristics, microprocessor data, and information on a variety of common support devices.

Microprocessors and microcomputers

Berk, A. A., *The Art of Micro Design*, Heinemann Professional Publishing 1980. ISBN 0 408 01 403 2.

A practically orientated guide to the design of microprocessor systems with particular emphasis on control applications.

Ferguson, John, *Microprocessor Systems Engineering*, Addison-Wesley. ISBN 0 201 14657 6.
Describes the basic skills, tools and techniques required to devise, develop and implement a microprocessor-based project.

Hall, Douglas, *Microprocessors and Digital Systems*, McGraw-Hill 1980. ISBN 0 07 025552 0.
Provides an excellent introduction to microprocessor-based systems and includes chapters on the use of test equipment, digital logic gate characteristics and interfacing, flip-flops counters and shift registers, D/A and A/D converters, microprocessor structure and programming and prototyping/troubleshooting microprocessor-based systems.

McGrindle, J. A., ed., *Microcomputer Handbook*, Collins. ISBN 0 00 383026 8.
A comprehensive guide to all facets of microcomputers. Includes sections on microcomputer boards and systems, software, development systems, and test equipment.

Putman, Byron W., *Digital and Microprocessor Electronics*, Prentice-Hall. ISBN 0 13 214354 2.
A comprehensive guide to the design and troubleshooting of modern digital and microprocessor-based systems.

Whitworth, Ian R., *16-Bit Microprocessors*, Collins 1984. ISBN 0 00 383113 2.
Provides a comprehensive introduction to 16-bit microprocessors and includes details of all popular 16-bit processor families. Chapters are also devoted to interfacing, instruction sets, assembly code software and development, system software and operating systems.

Programming

Brodie, Leo, *Starting FORTH*, Prentice-Hall 1982. ISBN 0 13 842930 8.
An entertaining and eminently readable guide to FORTH. Highly recommended.

Coffron, James, *Programming the 8086/8088*, Sybex 1983. ISBN 0 89588 120 9.
A useful introduction to the 8086 and 8088 for the assembly language programmer.

Denning, Adam, *C at a Glance*, Chapman and Hall/Methuen. ISBN 0 412 27140 0.

An excellent and reasonably priced introduction to C programming. Contains numerous examples.

Kernighan, Brian and Ritchie, Dennis, *The C Programming Guide*, Prentice-Hall 1978 ISBN 0 13 110163 3.

The definitive guide to C. An essential book for any aspiring C programmer.

Leventhal, Lance A., *Z80 Assembly Language Programming*, Osborne/McGraw-Hill 1981. ISBN 0931988 21 7.

A definitive work for those using Z80 assembly language. Each Z80 instruction is fully explained and more than eighty sample programming problems are discussed.

Leventhal, Lance A. and Saville, Winthrop, *Z80 Assembly Language Subroutines*, Osborne/McGraw-Hill, 1982. ISBN 0 07 931091 5.

Discusses Z80 programming methods and contains a large number of fully-documented routines.

Morgan, Christopher and Waite, Mitchell, *8086/8088 16-Bit Microprocessor Primer*, Byte/McGraw-Hill. ISBN 0 07 043109 4.

An excellent introduction to the characteristics, internal architecture and programming of the 8086, 8088 and associated support devices.

Purdum, Jack, *C Programming Guide*, Que. ISBN 0 88022 157 7.

An excellent tutorial guide eminently suitable for newcomers to the C language.

Purdum, Jack, *C Self-Study Guide*, Que. ISBN 0 88022 149 6.

A directed study guide with questions and answers. Useful as a supplement to Jack Purdum's *C Programming Guide*.

Williams, Steve, *Programming the 68000*, Sybex. ISBN 0 89588 133 0.

Provides a thorough introduction to the 68000, 68008 and 68010. Also contains information on the Unix and CPM 68K operating systems.

Index

Absolute address, 114
AC power failure, 7, 13
AC sensing, 199, 203
Accumulator, 26, 119, 126
ACFAIL*, 7
ACIA, 40
Active-high, 307
Active-low, 34, 35, 307
ADC, 76, 174
Address, 22, 23, 34, 307
Address bus, 4, 22, 307
Address decoder, 30, 35, 36, 307
Address decoding, 34
Address lines, 22, 51
Address range, 22
Address selection, 307
Address strobe, 52
Addressed board, 307
ADRSTB*, 52, 56, 57, 58, 59, 60, 65, 67
Alarms, 208
Alternate register, 21
ALU, 24, 26, 109
Analogue I/O, 75, 76, 174
Analogue I/O bus, 265
Angular position sensor, 176
Angular velocity sensor, 176
Arbiter, 49, 54, 300, 307
Arbitration, 12, 54, 307
Architecture, 10, 24
Arcom, 20
Arguments, 168
Arithmetic logic unit, 24, 26
ASCII, 42, 43, 110, 257
Assembler, 42
Assembler directive, 42
Assembly language, 42, 45, 106, 108, 109, 170
Asserted, 307
Asynchronous bus, 9
Asynchronous mode, 40
ATN, 17
ATNRQn*, 53, 59, 60, 63, 65, 68, 245
Attention line, 17
Attention request, 52, 60, 61

Audible outputs, 206, 207
AUXIN, 84
AUXOUT, 84

B, 46
Backplane, 9, 62, 307
Backplane bus, 5
Backplane signals, 256, 304
Backspace, 105
Back-up, 82
Banked memory, 80
Base pointer, 128
BASIC, 45, 143, 145, 271, 294
Basic disk operating system, 81
Basic input/output system, 81
Battery backed, 32
BCPL, 46
BDOS, 81, 96, 99, 100, 101, 106
Binary, 259
BIOS, 81, 96, 97, 100, 106
Bipolar RAM, 33
Bit, 307
Bit mapped, 33
BITBUS, 218, 219, 220
Block transfer, 55, 56, 308
Board, 308
Boot, 82
Boot loader, 81
BS 4937, 201
Buffer, 40
Burst mode, 55, 56, 60, 302
Bus, 4, 6, 9, 22, 308
Bus acknowledge, 53
Bus architecture, 9, 10
Bus connector, 67
Bus cycle, 52, 308
Bus driver, 30, 65
Bus extender, 253
Bus master, 6, 10
Bus processor, 69, 72
Bus receiver, 65
Bus request, 53
Bus slave, 6, 10
Bus termination, 65, 66, 69
Bus timeout, 245

Bus vectored interrupt system, 13, 14, 301
BUSAKn*, 53, 54, 55, 65, 67, 254
BUSRQn*, 49, 53, 54, 55, 65, 67, 254
Byte, 308

C, 46, 159
Capacitive proximity switch, 177, 179, 192
Card, 214, 215, 308
Carry flag, 27
CCP, 82
Central processing unit, 308
Chip, 308
Chip enable, 30
Chip select, 30, 34
Clock, 28, 29, 40, 69, 70
Clock cycle, 28, 29
Clock signal, 28, 40, 53
CMn, 52, 57, 65, 67
CMOS RAM, 32, 33, 215
Cold boot loader, 81
CONIN, 84
CONOUT, 84
Console command processor, 82
Console input, 100
Console output, 100
Constructs, 144, 152
Conversion characters, 164
Command files, 84
Command lines, 52
Command modifier, 52
Command tail, 84
Commands, 84
Comments, 140
Communications, 218
Compatibility, 308
Compiler, 43, 160
Condition code register, 26, 109
Conditional assembly, 113
Conflict, 30
Connector, 67, 70
Control bus, 5, 23, 27
Control characters, 86
Control key, 85, 86
Control logic, 24
Controller, 16, 49
Copying disks, 83
CP/M, 80, 87, 89, 99
CP/M-Plus, 79, 80, 82, 84, 86, 105, 114
CP/M-3, 79, 87
CPU, 245, 308
Cross reference utility, 110, 115

Cross-talk, 15
CTRL, 85

DAC, 76, 174
Daisy-chain, 308
Darlington transistor, 173, 205, 206
DART, 40
Data, 308
Data bus, 4, 22, 308
Data communication, 40
Data entry, 178
Data files, 157
Data lines, 52
Data strobe, 52
Data transfer acknowledge, 52
Data transfer bus cycle, 308
Data transfer protocol, 54
Data transfer sequence, 55
Data transmission, 40
DATACK*, 52, 56, 57, 58, 59, 60, 65, 68, 245, 254
DATSTB*, 52, 56, 57, 58, 59, 60, 65, 67, 245, 254
DC motors, 208
Debounce circuit, 183, 184
Debugger, 43
DEC, 18
Decoupling, 66
Default drive, 82
Default master, 49, 300
Delay, 120
Demultiplexer, 36
Destination disk, 84
Destination index, 128
Diagnostic tests, 245
Differential pressure/vacuum switch, 178
Diffuse scan proximity switch, 178
Digital I/O, 71, 73, 74, 172, 213
Digital I/O bus, 264
DIL switch, 178
DIN-41612, 17, 18, 19, 21, 77, 78, 303, 304
Directory, 84, 87
Disk controller, 214
Disk copying, 83
Disk directory, 84, 87
Disk files, 158, 171
Disk formatting, 83
Disk I/O, 41
DMA, 16, 18, 30, 61
Documentation, 139, 141, 142
Down-time, 244

Driver, 39, 206, 209, 211

EAROM, 32
Editor, 42, 110
EEROM, 32
Electromagnetic vibration sensor, 180
Enable, 30, 35
Encoded disk, 176
Encoded keyboard, 178
Encoded keypad, 178, 187
Environmental testing, 243
EPROM, 31, 32
Error checking, 139
Error conditions, 61
Error message, 43
Error trapping, 144
Eurocard, 1, 15, 19, 299
Explicit response, 62
Extrinsic commands, 85

Factory data communications, 218, 219
Fault-finding, 246
Fault location procedure, 251
Fault tolerance, 244
FID, 220
Field data messages, 219
FIFO, 60
File specification, 95, 110
Filename, 95
Filespec, 95
Filetype, 95, 96
Firmware, 308
Flag, 27, 121, 129
Flag register, 26, 109, 121, 129
Float switch, 177
Flow diagram, 8
Flow sensor, 177
Flowchart, 130
Fluid sensor, 196
Formatting disks, 83
Forth, 46
Functional elements, 48
Functional module, 308
Functional tests, 243
Functions, 151, 162

G-64, 17, 21
GESPAC, 17
GMT Electronics, 1
GPIB, 16, 17

Hand assembly, 45
Handshake, 308

Handshake lines, 38
Handshake protocol, 8, 10, 11, 54
Handshaking, 38
Hang, 244, 245
Hardware, 308
Hardware design, 221
Hardware hang, 244
Headers, 140
Hewlett-Packard, 16
Hexadecimal, 257
Hexadecimal loader, 45
High level language, 44
High state, 309

I/O, 24, 29, 36, 71, 74, 76, 144, 172, 309
I/O control, 37
I/O lines, 38
I/O port, 118, 172
I/O read, 58
I/O write, 58
IBM-PC, 20
IEC-603, 17, 67
IEC-625, 17
IEEE-488, 16, 17, 266
IEEE-696, 15, 16
IEEE-796, 18
IEEE-1000, 2, 19, 49, 54, 56, 63
Implicit response, 61
Index register, 123, 128
Indexed address, 123
Indicators, 203
Inductive load, 211
Inductive proximity switch, 179, 181
Industrial weighing station, 234
Input/output, 309
Input port, 309
Input validation, 139
Instruction cycle, 28, 29
Instruction decoder, 27
Instruction pointer, 26, 129
Instruction register, 27
Instruction set, 26
Integrated circuit, 309
Intel, 18
Intelligent slave, 6, 309
Interactive testing, 43, 116
Interboard signalling, 60
Interface, 309
Interface system, 309
Interfacing, 172
Internal bus, 4
Interpreter, 43

Interrupt, 13, 37, 132
Interrupt flag, 28
Interrupt handling, 61, 144, 301
Interrupt mask, 28, 132
Interrupt pointer, 134, 135
Interrupt request, 13, 27
Interrupt service routine, 28, 132
Interrupt vector, 123
Intrinsic commands, 85, 87

Joystick, 178

Key inputs, 153
Keyboard input, 41, 152, 178
Keypad, 178, 186

Labels, 111, 140
Languages, 44
Latching action switch, 185
LCD displays, 204
LED indicators, 203
Level shifter, 40
Library manager, 110, 115
LIFO, 27
Light dependent resistor, 177, 201
Light emitting diodes, 203, 204
Light level sensor, 177
Line numbers, 146, 147
Linear position sensor, 177, 193
Linear variable differential transformer, 177
Linear velocity sensor, 177
Linker, 43
Linking, 114
Linking loader, 107, 110, 114
Liquid level sensor, 177
List file, 114
Listener, 16
Load cell, 180
Loader, 43, 45, 82
Loading, 114
Local action response, 62
Logic levels, 65, 66
Logic probes, 249
Logic pulsers, 249
Logical constructs, 144, 152
Logical device, 84
Loops, 164
Low level language, 44
Low state, 309
LSI-11, 18
LVDT, 177

M-cycle, 28, 29
Machine code, 42, 43, 45, 144
Machine cycle, 28, 29
Macro, 111, 112, 113, 135, 136, 138, 139
Macro-assembler, 107, 110, 112
Macro library, 139
Magnetic sensor, 177
Mains connected loads, 210
Manufacturing automation protocol, 218
MAP, 218
Mask programmed ROM, 31
Master, 6, 10, 49, 309
Mean time before failure, 244
Memory, 30, 309
Memory block, 34
Memory card, 215
Memory controller, 33
Memory map, 23, 24, 81
Memory mapped I/O, 24, 36
Memory read, 58
Memory refresh, 33
Memory write, 58
Menu selection, 167
Messages, 152, 163
Microcomputer, 22
Microprocessor, 22, 24, 309
Microprocessor system, 29
Microsoft, 110
Microswitch, 178, 179, 190
MITS, 15
Mixed data, 55, 56
Mnemonic, 42
Motorola, 17, 19
Motors, 208, 211
MTBF, 244
Multibus, 18
Multiline interrupt system, 13, 14
Multiplexed bus, 5
Multiprocessor system, 10
Multirange meter, 247
Multitasking, 144

Names, 140
Negate flag, 27
Networking, 218
NMOS dynamic memory, 33
NMOS static memory, 33
Noise, 254, 255, 309
Noise margin, 66
Non-maskable interrupt, 28
Numeric variables, 147

Index 319

Numerical inputs, 156

Object code, 42, 109
Opcode, 111
Operating system, 41, 79, 309
Operation code, 111
Operator input, 178
Operator interface card, 215
Optical proximity switch, 179, 193
Optical receiver, 176
Optical transmitter, 176
Optically isolated inputs, 196, 217
Opto-isolator, 198
Oscilloscopes, 250
Output port, 309
Overflow flag, 27
Overload tests, 243

Page zero, 82, 100
Parallel, 36
Parallel I/O, 36, 37, 38, 39, 172, 173
Parallel arbitrated system, 12
Parameter passing, 132
Passing arguments, 168
Password, 95
PDP-11, 18
Peripheral, 309
Photocell, 177
Photodiode, 177
Phototransistor, 177
Physical device, 84
Physical machine, 41
PIA, 38
Piezo-resistive sensor, 179
PIO, 38
PISO, 37
Pixel, 33
PLA, 35
PLC, 1
Polled I/O, 37
Polled interrupt system, 13
Port, 38, 118, 120
Port I/O, 24, 36
Port initialization, 173
Position sensor, 176, 177, 193
Potential master, 10, 300
Power failure, 7, 13, 63
Power supplies, 63, 64, 77
Power-up, 54
PPI, 38, 74
Pressure sensor, 178
Priority, 12, 37, 61, 301
Procedures, 151, 294

Processor, 309
Processor cards, 216
Profile digitizer, 223
Program counter, 26, 28, 134
Program header, 140
Program status word, 26
Programmable logic controller, 1
Programmed I/O, 37
Programmed logic array, 35, 300
Programming languages, 44
PROM, 31
Prompts, 152, 165
Propagation delay, 9, 15
Protocol, 309
Prototyping card, 215
Proximity sensor, 179, 190
Pseudo-mnemonic, 42
Pseudo-operation code, 111
Push-button, 178, 181

Q-bus, 18
Quality procedures, 243

RAM, 24, 27, 29, 30, 32, 34, 310
RAM diagnostic tests, 246
Random access, 310
Random access memory, 310
Read, 26, 310
Read cycle, 301
Read operation, 22, 27
Read sequence, 57
Read-modify-write, 55, 56, 58, 302
Read-only memory, 310
Read/write line, 27, 34
Reed switch, 179, 190
Refresh, 33, 118
Refresh register, 124
Register, 24, 26, 109, 112, 118, 119, 124
Regulator, 15
Relative address, 114
Release, 310
Release-on-request, 49, 300
Release-when-done, 49, 300
Reliability, 243, 244
Relocatable module, 115
Reset, 28, 54, 82
Resident system extension, 82
Resistive strain gauge, 179
Restart, 82
Restart instruction, 133
Rollover, 187
ROM, 24, 29, 30, 34, 144, 215, 310

ROM diagnostic tests, 246
Rotary position sensor, 176, 194
Rotating vane flow sensor, 177
Rotational sensor, 176
RS-232C, 218, 268
RS-422, 218
RS-485, 218
RSX, 82
Rubout, 105
Run-time, 43
Run-time library, 43, 115

S-100 bus, 15
Screen memory, 33
Secondary bus, 302
Segment register, 125, 126, 129
Semiconductor strain gauge, 179
Semiconductor temperature sensor, 180, 199
Sensors, 175, 180
Sequence, 310
Sequence diagram, 7, 11
Serial, 36
Serial arbitrated system, 12
Serial I/O, 36, 37, 40, 41
Settling time, 310
Shaft encoder, 176, 194
Shift register, 37
Signal, 310
Signal level, 310
Signal line, 51, 310
Signal parameter, 310
Signal voltages, 65, 66
Single data, 54, 56
SIPO, 37
Slave, 6, 51, 310
Snubber, 211
Software design, 222
Software development, 79
Software hang, 244
Software tools, 79, 110
Solenoids, 210
Solid-state relays, 210
Source code, 42, 43, 109, 111, 159
Source disk, 84
Source index, 128
Specifications, 220
SSDA, 40
Stack, 27, 33, 134
Stack pointer, 27, 122, 128
Standard input, 165
Status indicators, 203
Status register, 26, 109

STD, 1, 17, 21
STE, 1, 19, 21
STE Bus Product Guide, 263
STE Manufacturers and Users Group, 270, 296
Stepper motors, 211
Stock faults, 247
Strain sensor, 179
String inputs, 156
String variables, 147
Structured English, 130
Structured programming, 44, 130
Subroutine, 120, 132, 137, 149
Supervisor stack pointer, 27
Supplies, 63
Switch, 178, 180
Switch debouncing, 181, 183, 185
Symbol table, 113
Symbolic address, 42
Symbolic debugger, 107, 116
Symbols, 111, 140
Synchronous bus, 9
SYSCLK, 49, 65, 68, 254
SYSRST*, 7, 49, 65, 68, 244, 245, 252
System, 310
System clock, 49, 53
System control block, 96, 105
System controller, 49, 54, 60
System loader, 81
System reset, 7, 49, 53
System software, 42
System stack pointer, 27
System timeout, 60
System variable, 33

T-state, 28, 29
Tachogenerator, 176
Tachometer, 176
Talker, 16
Talker/listener, 16
Temperature sensor, 179
Tenure, 310
Terminal, 77
Terminal emulation, 77
Termination networks, 65, 66, 69
Test equipment, 247
Testing, 139, 243
Text editor, 110
Text file, 42, 43
Thermistor, 179
Thermocouple, 179, 200
Threshold detection, 201
Timeout, 245

Timing diagram, 7, 11
Toothed rotor tachometer, 176
Top-down approach, 220
Touch operated switch, 185
TPA, 82
Transaction, 310
Transfer, 310
Transfer error, 60
Transient program area, 82
Trenan, 219, 220
TRFERR*, 49, 52, 60, 65, 68, 245, 254
Tristate, 30

UART, 40
Ultra-violet, 31
Unencoded keyboard, 178
Unencoded keypad, 178, 187
Unibus, 18
Up-time, 244
USART, 40
User defined functions, 151
User stack pointer, 27
Utility commands, 89
Utility lines, 53

Utility programs, 79, 89, 110

Validation, 139
Variables, 147
VDU, 84, 310
Vector, 14, 28, 123
Vector fetch, 52, 55, 59
Versabus, 19
VIA, 38
Vibration sensor, 180
Virtual machine, 41
Visual display unit, 310
VME, 19, 298, 303
VMOS FET, 205, 206, 208, 209

Watchdog controller, 63, 244, 245
Weight sensor, 180
Write, 26, 310
Write cycle, 301
Write operation, 22, 27
Write sequence, 57

Zero flag, 27
Zero page, 82

Index to microprocessors and VLSI support devices

2114, 34
2128, 34
2516, 32
2532, 32
2564, 32
2716, 32
2732, 32
2764, 32
4116, 34
4118, 34
4164, 34
4256, 34
4464, 34
4864, 34
6116, 34
6264, 34
6502, 23, 27, 28, 109
6520, 38
6521, 38
6522, 38
6809, 17, 27
6820, 38
6821, 38

6850, 40
6852, 40
8044, 220
8080, 15, 80, 108, 116, 118, 124, 133
8085, 25, 28, 80, 108, 124
8086, 80, 108, 124, 134, 135
8088, 108, 124
8251, 40
8255, 38, 74, 173
8256, 40
8291, 17
8292, 17
27128, 32
27256, 32
27512, 32
41256, 34
43256, 34
50464, 34
68000, 19, 27, 80, 109
68488, 17
Z80, 24, 27, 29, 34, 69, 80, 108, 111, 116, 118, 119, 124, 130, 133
Z80-DART, 40
Z80-PIO, 38